GW01325876

The First Time I Rode a Freight Train
& other hitchhiking stories

By Tim Shey

PublishAmerica
Baltimore

First printing

Hardcover 9781462661718
PUBLISHED BY PUBLISHAMERICA, LLLP
www.publishamerica.com
Baltimore

Printed in the United States of America

Author's Note

This is a collection of short stories and other writings about my hitchhiking experiences in the United States. Most of these stories were published by *Sawman* (Tim Shey) on *Digihitch.com.* The short story "High Plains Drifter" and the poem "Shiloh" were first published by *Ethos*, Iowa State University, Ames, Iowa. The letter to the editor "Why is Hitchhiking Illegal in Wyoming?" was originally published by the *JH Weekly*, Jackson, Wyoming. The poem "Egypt is Burning" and a few other pieces were first published on my blog *High Plains Drifter* .

All Bible quotes are taken from the King James Version

***High Plains Drifter**:* http://tim-shey.blogspot.com/
***Digihitch Road Profile**:* http://www.digihitch.com/account_Sawman.html
***The Highway**:* http://www.digihitch.com/blog/sawman/index.html

The First Time I Rode a Freight Train

Back in July of 1980, I was house-sitting for some friends in Ames, Iowa. They and their two daughters were gone for a month or so seeing relatives in Southern California.

One day I decided to hit the road and see how far west I could get. I took my backpack and some of my belongings and began hitchhiking west on U.S. 30.

I got a few rides to Denison. Then this young lady picked me up near Dow City or Dunlap. She had a can of beer in her hand and offered me one; I declined the offer. I went to enough beer parties in high school; my beer-drinking days were pretty much over. She was fairly drunk and she would swerve over into the other lane every so often and then correct herself.

Finally, I said, "Hey, if you want, I can drive for you."

She said, "No. I'm doing just fine."

A few minutes later she barely missed hitting this tractor-trailer coming from the opposite direction.

I had had enough, so I said, "Pull over and let me out."

She pulled over onto the shoulder and I got out of the car. She gave me the finger and drove off. I was so glad to get out of that vehicle. That was the first time (and maybe the only time) I asked to get out of a car because the driver was drunk.

So I walked down the road and this guy picked me up. He had just graduated from the veterinary school at Iowa State University in Ames. This guy was going to Nebraska to take his boards for the state of the Nebraska.

He dropped me off someplace and I later made it to Blair, Nebraska. The sun was setting as I walked down main street. I walked past this gas station and this kid that worked there was sitting in a chair.

He looked at my backpack and asked, "Where ya goin'?"

"I'm heading out west," I replied.

"Have a good trip."

"Thanks."

I walked through Blair and was a mile or so out of town, when a sheriff deputy stopped me. He asked me where I was going and then he checked my ID. At the time, I was a little annoyed that they would stop and check me. I was walking down the road minding my own business. What's the big deal, I thought. That was probably the first time I had been stopped by law enforcement for walking or hitchhiking. I was a little rattled about the whole thing.

The sheriff deputy gave me my ID back and I continued walking due west on U.S. 30.

The sun was down, so I decided to jump over this fence and hightail it to the railroad tracks. It wasn't long and I was walking down the tracks of the Union Pacific.

I had been walking for a while when all of a sudden this powerful light came around the bend behind me and this locomotive was bearing down on me! I didn't even hear it coming! I took evasive action, quickly jumped off the tracks and ran into the ditch. The four or five engines roared past with its grain cars in tow. That was a close one, I thought.

I later learned that the sound of the engine travels out from the sides of the locomotive, not from the front. I had been hitchhiking in New Mexico back in the late 1990s, when this man and his wife and kids picked me up. He worked as a welder for the Santa Fe Railroad. He told me about the sound traveling out from the sides and not from the front. He and his fellow welder almost got run over by a train while they were welding "frogs" on the tracks. They never heard the train coming--just like in my case.

So I continued to walk down the tracks. I then camped out in some grass. It was hot and humid--it was probably in the upper nineties that day. The mosquitoes were bad. I don't think I got much sleep that night.

The next morning I got a couple of rides to Fremont.

I was walking in downtown Fremont heading towards the railroad tracks (I was thinking about hopping a freight train) when a local cop stopped me.

"Where you going, son?" he asked.

"I'm heading west," I said.

"You're going in the wrong direction. Hop in and I'll give you a ride west of town."

"Sounds good."

He dropped me off near this pond of water; it looked like a state park or campsite. I thanked him for the ride and he drove off.

It was now around a hundred degrees and I was getting hot, so I spent some time swimming in the pond. After a while, I lay down on this picnic table and took a nap for an hour or two.

Then I heard this low rumbling. I woke up and saw this freight train slowly moving westward on the tracks maybe a hundred yards away. I quickly put on my socks and boots and grabbed my backpack and ran to this brand, spanking-new flatcar. It had my name written all over it.

I climbed onto the flatcar and put my backpack against the bulkhead. I sat down and rested my back against my backpack. The train merged from the siding onto the main line and gained some speed. I was now in business.

It was exhilarating and free, sitting on that flatcar watching the green Nebraska countryside go past. Eventually, I took off my boots and socks and sat on the flatcar barefoot. I felt even more free. The train was now traveling at around fifty miles per hour.

The Union Pacific tracks ran parallel with U.S. 30. Cars and pickups would drive down the highway and people would wave at me and laugh. I would wave back and smile. Some people would honk their horns. It was a lot of fun.

The train rolled through North Bend and Schuyler and finally slowed down and stopped in Columbus. We weren't stopped very long. They were switching out some cars, I'm guessing.

The train slowly moved out and we were heading west again.

My plan, when I left Ames, was to see some relatives in Ogallala, Nebraska. Well, I didn't know their names and I didn't know exactly where they lived in Ogallala. I just knew that I had some relatives in Ogallala and this is why I headed west. It doesn't make a whole lot of sense now, but back then I was twenty years old and all I wanted was an excuse to hit the road and head west. Relatives in Ogallala: sounds good to me. (I later did meet these relatives in Ogallala back in 1983--just before I hitchhiked to California for the first time.)

The train was now going down the tracks at a pretty good clip. I was absolutely enjoying everything about life on a flatcar when I saw this Nebraska Highway Patrol drive by on U.S. 30. I smiled and waved at him, but he didn't wave back. He gave me a dirty look. It was then that I began to think that maybe I wasn't supposed to be riding this freight train.

I didn't think it was illegal to hop freight trains (but that maybe some people might frown on it). My great-grandfather, who was born in County Roscommon in Ireland, lived for thirteen years in Australia herding sheep and prospecting for gold. He came to America and settled down in southwest Iowa. He used to ride freight trains between Iowa and western Nebraska all the time. But that was back in the late 1800s and early 1900s. I began to think that sitting on this flatcar in plain sight of everybody was not such a good idea.

The train rolled through Central City and soon began to slow down. As the train slowly made its way through the small town of Chapman, there was this cop in his car stopped at the intersection. I smiled and waved at him, but he didn't wave back. Then I began to get this sinking feeling. Maybe I better get off of this train ASAP.

Well, the train stopped maybe a quarter of a mile from where the cop was sitting. I saw the cop car back up and drive down the service road that ran parallel with the tracks. He stopped his car next to my flatcar and motioned for me to get off the train.

I looked at him and said, "Who me?"

He nodded his head as if to say, "Yes, you."

I put my socks and boots back on and climbed off of the flatcar. I didn't like this one bit.

I got in the car and the cop told me that it was illegal to ride freight trains. He drove me to Central City to the police station. He said that he was going to contact the "U.P." (Union Pacific) detectives and see if they wanted to prosecute me.

I sat at the police station as the officer phoned the Union Pacific. The guys in the jail asked me why I was there. I told them that I got caught riding a freight train. They howled in laughter. I sort of laughed, but not really.

The officer hung up the phone and told me that the Union Pacific didn't want to prosecute. He told me to get back in the car and that he would drive me east to the county line.

As we drove east on U.S. 30, the cop asked, "So Tim, do you ever think about where you will go when you die?"

I answered, "Yeah, I think about it all the time."

So he began to tell me about Jesus and the Gospel. We had an intense talk. I was not yet a Christian, but this cop definitely sowed some good seeds into me. I asked Christ into my life two years later. Getting caught on a freight train by a Christian cop was definitely the hand of God--but I didn't know it at the time.

I was restless and seeking something: truth, beauty, literary aspirations, freedom from Adamic slavery. I dropped out of high school twice because it was so oppressive and unchallenging. I was hungry and desperate. Heaven was on my mind. I was looking for God, but did not know how to truly access Him. In July 1980, I was not far from the Kingdom of Heaven.

The cop dropped me off in the middle of somewhere. It was ten o'clock at night, it was hot and humid and I forgot to fill up my water bottle back in Central City. I was not a happy camper. I thanked the officer for the ride and he turned around and drove west into the Nebraska night.

The next town was six miles away. So I walked past the corn fields and the hay fields of eastern Nebraska. I was thirsty. The noise of diesel engines roaring away pumping water into irrigation circles could be heard as I walked back east.

Eventually, I made it to the small town of Duncan. I found a water hydrant and drank a ton of water. I then found a pickup parked next to the railroad tracks. I climbed into the cab of the pickup and slept there that night.

The next morning, I walked to the shoulder of U.S. 30 and began thumbing for a ride to Columbus. Within half an hour, some guy walked up to the pickup that I had slept in the night before and drove off in it. Sometimes it is a good idea to get up early in the morning.

I got a ride to Columbus. This guy took me to the bus station. I met a lady there that helped me pay for a bus ticket to Des Moines. I got on the bus and it went through Omaha. I got off in Adel, Iowa that evening. Adel is just west of Des Moines on U.S. 6.

I phoned a friend in Ames. He picked me up in Adel and drove me back to Ames. He thought that it was funny that I hitchhiked to Nebraska and hopped a freight train. He thought it was really funny that a cop told me to get off the train. I didn't think it was so funny.

A Conversation with a World War II U.S. Navy Frogman

I believe it was back in September of 1999, when I was walking north on U.S. 95 somewhere near Beatty, Nevada that this older guy picked me up. He looked like he was in his 70s. He was coming from Mexico and going back to Northern California where he made his home. He told me that he was a Navy Frogman in World War II.

As a Navy Frogman, he would go onto an enemy beach at night and prepare it for a Marine amphibious assault. They would cut barbed wire, take out mines, get rid of enemy infrastructure and so on. One time he and his fellow Frogmen were trying to defuse a mine in the ocean and the mine exploded. I guess several Frogmen were killed; he and another guy survived.

He said that after the war, he did a job as a mercenary somewhere in Central America. He got caught by the local government or warlord and was thrown in prison. He heard a man screaming because he was probably being tortured--and, he thought, being killed.

I don't remember how long he spent in that prison, but he told me that he thought he was a goner. Then one evening something profound happened. He had an intense spiritual experience: he saw a vision of Jesus and this overwhelming sense of peace came over him. A few days later, he was released from prison and he did no more mercenary work after that.

He spent twenty or thirty years in the merchant marine as a cook. He had been retired for some time. His intestines were shot, so that is why he wore a bag on his side. He was married and divorced from an exotic dancer. His son was thrown in prison for robbing convenience stores. He seemed pretty wore out from living on the planet.

He told me something interesting. He said that whenever you go to a bar at a naval base where Marines and Navy personnel hang out, if you see a guy sitting at the bar drinking by himself, it is usually a Navy SEAL. So I asked him why. He said that you go through hell to become a Navy SEAL and so it separates you from the rest of the crowd. Also, he said that SEALs go on top-secret missions that nobody can know about, so they can't talk to anyone about their work. So who can they talk with?

It is lonely at the top.

I was hitchhiking in Iowa back in 1986 and I was talking with this guy about the Marines (I enjoy reading military history). Then he told me about the SEALs that they are the best-trained warriors in the world. He told about these four Marines that were sitting at a table in a bar and they were drunk and obnoxious and trying to pick a fight with somebody. Then this guy walked in and sat down at the bar and drank a beer quietly by himself. The Marines began making fun of him--they were trying to provoke something. The bartender walked over to the table of Marines and told them that the guy at the bar was a SEAL. The Marines quickly left the bar and never looked back.

"At the rebuke of His presence they fled." "A quiet word breaketh a bone." "The idols of Egypt are removed at his presence."

I guess you can say that that Navy SEAL's reputation preceded him.

A Hot Meal at a Campfire in Montana

I believe it was in January of 2002 when I got dropped off at the Flying-J Truck stop on the east side of Billings, Montana. I began to walk east on I-90 and walked past the intersection of I-94 and I-90. I continued walking on I-90 up that big hill due south. I probably walked several miles and got into ranch country. By now it was close to sundown.

I noticed this big culvert that ran underneath the interstate, so I walked down into the ditch and put my backpack in the culvert. I found a water tank nearby and walked to the tank and there was a hydrant, so I filled up my water bottle.

I gathered some sticks and anything that would burn and made a fire in the culvert. By now it was dark, so the light of the fire could be seen by anyone driving by--and it threw off some good heat, too. I think it got down to 27 degrees F that night, so it was good to get warmed by the fire.

Within a half hour this pickup pulled off the gravel road a hundred yards away and drove down to the culvert. This man got out of the pickup and asked, "Hey, what's going on?"

"I'm just passing through. Heading south tomorrow," I replied.

"Sounds good." He walked back to his pickup and drove off.

A half hour later, that same pickup drove back down to the culvert. Two men climbed out of the pickup and walked to my fire; one man was holding a plate of food in his hand.

"We thought maybe you could use something to eat," the older man said.

"Hey, thanks," I said. I was very grateful.

We spoke for a while warming ourselves at the fire. I began eating my hot supper--it really hit the spot. The older man was the father of the younger man. They had a ranch up the road.

I then said, "The Lord really knows how to provide."

The older man just shook his head and smiled. They stayed for a few more minutes and then walked back to their pickup and drove off into the night.

After my meal, I found a piece of plastic styrofoam and laid it on the concrete and rolled out my sleeping bag on top of it. It is very difficult to sleep on a slab of cold concrete--there needs to be some insulation between your body and the cold concrete. I remember I tried to sleep under this bridge on I-90 east of Butte, Montana one March or April, but I didn't have any insulation on the ice-cold concrete: I didn't sleep at all that night. We learn through experience.

In that culvert, I had a fire and some insulation to sleep on and I had a hot supper, so I slept well that night.

The next day I headed south into Wyoming.

A Providential Ride to Manhattan, Kansas

Back in 2001, I was walking from I-70 towards Manhattan, Kansas when this pickup pulled over to give me a ride. I climbed into the pickup and this young lady said, "The Lord told me to pick you up."

I said, "Praise the Lord! I'm a believer, too." So we had a great chat all the way into Manhattan.

So I asked her, "Were you raised in Kansas?"

"No," she replied. "I was raised in Indiana."

"Have you ever heard of Columbus, Indiana?"

"Yeah."

"Have you ever heard of Hope, Indiana?"

"Yeah."

"Have you ever heard of the St. Louis Crossing Independent Methodist Church?"

She looked at me and exclaimed, "My grandparents go to that church!"

"I was hitchhiking through Indiana a few years ago and this teenage kid picked me up and took me to his home in Indianapolis. I met his parents; they were very friendly people. His dad was the pastor of the St. Louis Crossing Independent Methodist Church; he asked me if I wanted to give a message at the church the next day. I said that that would be great. I gave a message at their church on Sunday and your grandparents probably heard me speak."

"That's incredible."

The young lady told me of the time she was driving through Kansas a couple of years earlier. She saw this older man hitchhiking on the side of the road. In the goodness of her heart she wanted to give him a ride, but the Lord said, NO. So she drove on by. Two days later, she saw that hitchhiker on the nightly news: he had robbed and killed an older couple in the next town.

It pays to obey the Lord.

I was hitchhiking in California a couple of summers ago and these two ladies picked me up. They were Christians and we had a nice talk. The one lady told me that she picked up this hitchhiker in the Bakersfield area and gave him a ride. He was very strange: he kept staring at her.

He said, "Aren't you afraid of me?"

She said, "No, I'm washed in the precious Blood of Jesus Christ!"

Immediately the hitchhiker froze up and didn't say a word. The lady dropped him off at the next town.

A few weeks later the lady saw that hitchhiker on the nightly news: he was in jail for killing three women in the Bakersfield area. In the interview, the hitchhiker said that he kidnapped this woman and took her out to the Tehachapi Desert and tried to kill her, but couldn't. I believe someone was praying for her safety.

When hitchhiking or picking up hitchhikers, put God first.

Psalm 91 is a great psalm to read and build up your faith; it is the psalm of protection.

A Christian Cult

This is a story about an experience I had at a Christian cult.

I was hitchhiking in eastern Pennsylvania back in April 1998 when this man and his son picked me up. When I got in the car, immediately, this guy gave me the creeps. I noticed a Bible on the dashboard, so I thought maybe he was a Christian.

We got to talking and he asked me why I was hitchhiking. I told him that I was hitchhiking from Iowa to New York; I wanted to visit David Wilkerson's church at 51st and Broadway in Manhattan. I am a Christian and love to read the Bible, but when we would talk about Scripture and the things of God, everything this guy said didn't seem right--it didn't bear witness with my spirit.

So he said that he lived in a Christian community in Saugerties, New York--in the Catskill Mountains. He said that it would be all right if I wanted to stay the night and then hit the road the next day. It was okay by me.

We drove to Saugerties to the "Christian community." There were several buildings at the compound. There were probably 50 men, women and children. He showed me around the place and then this other guy showed me more of the place. I didn't like being there right away.

These little kids would walk up to me like little robots and say woodenly, "Welcome to our community. It is so nice that you are here." Or words to that effect. It seemed like little canned speeches. I would meet adults and they would say the same thing.

It was soon late afternoon and they had an evening fellowship meeting in the main building. It lasted maybe half an hour: singing, dancing, a little message from one of the leaders--the Spirit of Christianity was not there. It seemed so fake and artificial. The Presence of God was not there at all. It reminded me of a church back in Ames, Iowa that I used to go to back in 1987-1988--Great Commission Church--very cultish.

For some reason, I noticed this guy on the other side of the room. You could tell that he hated being there: he would stay for a few minutes and then he would leave--he did this a few times.

After the meeting, we went to the building where everybody dined. I sat down at this table and the guy who I noticed at the meeting sat across from me. He knew that I was the new guy in town, so he asked me a lot of questions about hitchhiking. He had never gone hitchhiking before, so I told him to put his trust in God and that the Lord would protect him on the road. He finished his meal before everybody else and shook my hand and shook my hand and shook my hand and smiled at me and thanked me for telling him about hitchhiking. He walked out of the building and on purpose walked by the window where I could see him and he smiled at me. At the time, I wondered why I was seeing this.

After supper some of us sat down in the living room and I was surrounded by five people. They told me that they were the ONLY ONES doing the will of God and that I should stay there permanently. I told them that I had planned on staying one night and then I would head to New York City. I listened to their propaganda for at least three hours and then I went to bed.

I didn't sleep at all that night: I was on pins and needles. I told the Lord that I was in the pit of Hell and why did He put me here? I knew that God has everything under control, but I still wondered why I was at that cult.

Well, the next morning we all went to the morning fellowship meeting. The first thing that one of the elders said was, "Brother So-and-So escaped last night."

I stood there semi-shocked and said to myself, "Praise the Lord! That guy that I talked to last night escaped. Now I got to get out of here." Why would that elder use the word 'escape'? Man, what a flogging idiot.

So after the morning meeting, I walked back to my dorm room and packed up my bag and walked out the front door.

The guy who picked me up in Pennsylvania was waiting for me outside and started yelling at me, "Tim, what are you doing?!" He looked very concerned as if something was wrong with me.

I smiled at him and said, "I'm hitting the road just like I said."

Then he took this piece of paper and read what was written on it and said, "No. I am your leader. You must obey me. We are the ONLY ONES doing the will of God. You must stay here."

I said, "Well, you boys keep up the good work, but I'm hitting the road." I even shook his hand; it was like shaking a limp piece of nothing. He was really ticked off that I was leaving.

So I walked on down the road and eventually got rides all the way to New Hamburg, New York where I took a train into Manhattan.

I was so glad to get out of that Christian cult and I was grateful that the Lord used me to help that guy escape that place. But then I told the Lord, "Please don't ever do that to me again."

"Not my will, but Thine be done." "Obedience is better than sacrifice."

Maybe the title of this entry should be "Escape from New York."

The Only Time Someone Pulled a Knife on Me

Back in the late 1990s, I was hitchhiking through Cloudcroft, New Mexico and this pickup pulled over to give me a ride. There were two men in the front seat and they told me to hop in the back of the pickup. I noticed the eyes of the passenger: they looked crazy--like he was on drugs or something.

So I hopped in the back and they drove me to the next town to a trailer park. I hopped out of the pickup and began talking with the driver. He was a nice guy and we had a friendly chat.

Then I noticed that the passenger walked around to my right and began walking towards me. He then whipped out this knife (or a tool with a blade on it) and he lunged at me.

I quickly jumped back and said, “All right, are you guys trying to rob me or what?”

The driver of the pickup exclaimed, “No! This guy is an idiot! Throw down that knife, you idiot!”

The passenger threw down his knife.

The driver felt bad that his friend had pulled a knife on me and asked, “Hey, can I make you some lunch.” He pointed towards his trailer home.

“I think I better mosey on down the road,” I replied. I thought maybe they might be leading me into a bigger ambush.

“Come on inside and let me introduce you to my two daughters,” he said, as he walked to the trailer.

So I followed him inside and met his two daughters; they were around eleven and twelve years old. He asked me again if he could make me some lunch. I declined the offer.

Then he noticed my baseball cap which had “Harold Pike Construction Company” written on it. “Hey, can I have your cap?” he asked.

"No problem," I answered. We exchanged caps. His cap had "Indiana University" written on it.

He asked again if he could make me some lunch. I said, no, that I better hit the road. He gave me a few bucks. He told me that his name was Apache; he also gave me a Gideon's New Testament. As I walked out the door of the trailer, the guy who pulled the knife on me gave me a dollar bill and I shook his hand.

"No weapon formed against thee shall prosper."

Psalm 146: 9: "The LORD preserveth the strangers; he relieveth the fatherless and the widows: but the way of the wicked he turneth upside down."

Sleeping at the Post Office in Bridgeport, California

I am guessing in December of 2006, I was hitchhiking up U.S. 395 from someplace--maybe Ridgecrest, California--and heading north to Reno. I ended up in Bridgeport that evening.

I walked around Bridgeport for a while. I was looking for a barn or an abandoned car to crawl into to keep warm. It was going to get down below 20 degrees F that night. Finally, I walked over to the post office and put my backpack in the corner and sat on it for a while--I was cold and tired.

I had been sitting on my backpack for maybe half an hour, when this cop walked into the post office. He saw me sitting there and said, "Well, it's going to get cold tonight. This is probably the best place for you to sleep."

I said, "Thanks."

I was grateful that he didn't kick me out of the post office. I had been kicked out of post offices in South Dakota and Nebraska. I have slept in several post offices in my hitchhiking travels.

So the next day, I packed up my things and moseyed toward Reno.

Six months later, during the summer of 2007, I was again hitchhiking up U.S. 395. I got a ride from just north of Lee Vining to Bridgeport. It was a California Highway Patrolman. He was friendly and we had a nice chat.

He told me that he lived in Bridgeport. He said that sometime last winter, he walked into the post office and there was this guy sitting on his backpack.

"That was me!" I exclaimed.

"So you're the guy!" He started laughing.

"Thanks for not kicking me out of the post office. It was cold that night."

"No problem. The next time you come through Bridgeport during the winter, just go to the Sheriff's Department. They will let you sleep on a cot in the lobby."

And some people think that the post office is just for mailing letters.

My Backpack

I think I should write a little about my backpack. I carry it everywhere I go; it has been invaluable in my hitchhiking journeys.

Back in September of 1999 I was hitchhiking in northern California--somewhere on U.S. 395 north of Susanville--and this guy picked me up and asked me if I would help him do some carpenter work. I agreed and worked for him for about three hours.

After we were finished, he said he would give me his backpack because he didn't have any money to give me. I gave him my bag that I carried on my shoulder, took his backpack and put my stuff in it and have had it ever since. It was a real blessing because the backpack's weight is better distributed on your shoulders and back and hips than the bag that I carried on my shoulder--and I can carry heavier loads. I think my backpack has weighed up to forty pounds.

The things that I carry in my backpack are: a U.S. Army sleeping bag; a water bottle; a zippered folder that holds my manuscripts, CDs, a floppy disc, pens, address book, an atlas of North America and other papers; clothing; a shaving kit; batteries for my flashlight; a little all-purpose tool; toilet paper; moist towelettes; a little Gideon's New Testament; a pocket atlas of the United States; a King James Compact Reference Bible; some disposable Gillette razors; a plastic carrying case for six mini-CDs; two stocking caps; a small roll of duct tape.

My backpack has shown a lot of wear and tear over the years. There are rips in it; it is somewhat dirty. There are places where I sewed it up with monofilament fishing line and there is a piece of duct tape on the bottom of the pack. Without duct tape, we would be a people no more.

I believe the weight of my backpack averages around thirty-five pounds, so I get some good exercise every day when I have to walk several miles on the highway. The guy who gave me the backpack told me that he spent $200.00 for it back in 1979. It is still hanging in there pretty tough. It is an interior frame backpack. I don't know the brand name.

It has been through rain, snow, dirt, mud, sand (e.g. I slept on the beach at Cambria, California), crude oil (in the back of a pickup in New Mexico), hundred-degree heat, and twenty-below-zero cold. I use it as a pillow when I sleep outside. I use it as body armor when somebody drives by and sprays me with submachine gun bullets (just joking). My body armor is a wall of fire that surrounds me--the Holy Ghost Fire.

My backpack and I have hitchhiked countless thousands upon thousands of miles throughout the United States. Somebody once offered to buy me a new backpack two or three years ago. I graciously declined their offer. I'm going to keep this backpack as long as I can. You see, it never argues with me, it never disagrees with me, never talks back. It is very low maintenance. When I get tired of carrying it, I stop, take off my backpack and sit on it on the side of the road and rest for a while.

When I die, it doesn't look like I will be able to take it to heaven with me--I guess this is something that I will just have to accept.

A backpack, a backpack, my kingdom for a backpack.

Without a backpack, I would be a hitchhiker no more--or just another hitchhiker without a backpack.

The first backpack was probably invented somewhere between Cain and Abel and the time of Noah. The first hitchhikers probably came about just after the Tower of Babel: the Lord confused the languages of the people and the people were forced to migrate to the four corners of the known world, so there must've been a lot of people looking for rides on oxen-driven carts and on camel caravans.

I have heard that U.S. Marines carry eighty-pound backpacks in boot camp and that British SAS (Special Air Services) men carried two hundred-pound packs in Operation Desert Storm (1991). Thirty-five pounds doesn't feel so bad. It's my backpack and it doesn't complain: I'll keep it as long as it holds up.

Washing Dishes

A year or two ago I was hitchhiking across the Navajo Indian Reservation in northeast Arizona and I got a couple of rides to Flagstaff. It was during the winter and it was going to get cold that night (maybe around 0 degrees F), so I stayed at a Christian mission in downtown Flagstaff.

They have a well-run mission there. After 5 PM, one of the leaders would give a Gospel message and then we would have supper. After supper, we would shower and then go to bed. They have a dorm room upstairs; I believe they have beds for twenty men.

So the next morning we were eating our breakfast and one of the leaders asked everybody, "So who wants to volunteer to wash dishes?"

Immediately, I raised my right hand and said that I could wash the dishes. The leader smiled at me, walked over to me and patted me on the back.

Then the leader asked, "Who wants to help Tim wash the dishes?"

Nobody raised their hand.

The leader looked at this guy and asked, "Hey, Hank, why don't you help Tim wash the dishes."

Hank replied with a look of disgust, "Now that is not a Christ-like thing to say." Which meant he didn't want to wash the dishes.

So the leader said, "Well, Hank, if you don't want to wash the dishes then go back outside." And Hank left the mission.

I just about couldn't believe what I had heard. Washing the dishes is a very simple, easy job. And your hands get cleaned in the process. Hank got a free meal and couldn't wash the dishes. Ingratitude comes in different wrappers.

I was very grateful that that Christian mission let me stay there out of the cold for one night. They preached a good message the evening before, I had a great supper, I was able to take a shower and sleep in a warm bed and then have an excellent breakfast the next morning. If someone wants me to wash the dishes, then I'll wash dishes till the cows come home!

Once I was hitchhiking through Pennsylvania and this guy picked me up. He had a used auto dealership and asked me if I wanted to help drive a car from one town to the next. I said no problem. Then he said, let's go to this mission and get some lunch. He usually recruited guys from that mission to drive cars for him.

So we signed in at this mission--I believe it was in York, Pennsylvania. I was the last guy in line and the guy ahead of me was definitely a street person. He had a real bad attitude. He kept complaining about the food: "I don't like this crap. Why do I have eat this junk? Don't you guys know how to cook a meal?" And words to that effect.

So I went through the line and thanked everyone for the great meal and smiled at everyone. Redemption sometimes happens in soup lines.

That street person didn't pay for his meal, didn't prepare it, didn't volunteer to help in anyway, but he sure complained to everyone there about the food. Then go outside and eat grass!

Nobody there asked me to help wash the dishes, so I hung out with the used auto guy for a while and then moseyed out west on U.S 30.

A Christmas Story or Junked Cars Can Be Beautiful

I hitchhiked from Bozeman to Big Sky [Montana] yesterday afternoon. When I got to Big Sky it was 4:30 P.M. I walked south a few miles and soon it was nightfall.

I walked past this restaurant/bar and saw this junked car to my right. I walked up the slope to the car and it was covered with snow. I crawled inside the back and rearranged some things that were stored in it so that I could make room for me and my sleeping bag. Well, somebody who worked at the restaurant/bar saw me and told me to get out of the car; he said that I would freeze to death--it was too cold. So I rearranged what I had rearranged in the back seat of the car, hefted up my backpack onto my shoulders and made my way south down the moonlit highway towards West Yellowstone. I was complaining a little bit: I didn't know why I had to hitchhike at night in the dead of winter in a snow-covered canyon. I knew that I was there for a reason, so I wasn't worried or all bent out of shape about the whole situation: I knew that the Lord would not leave me stranded forty miles from nowhere when it was that cold.

Eventually, I did get a ride with a guy who was going all the way to Idaho Falls. He was driving a pickup and had his two dogs sitting in the cab with him. I was very grateful that he picked me up. The road was pretty icy going towards West Yellowstone. We got to West around 8 P.M. It was 10 degrees F. We stopped at a gas station and I kicked him down five bucks for gas and I got a hot chocolate and corn chips for the road. We continued south and the roads were still snowy and icy till we got south of Last Chance/Island Park.

As we drove through Island Park, he told me that some local Nazis burned down his log cabin (he used to live in Island Park) because he didn't subscribe to their philosophy. So now he lived up in the Bridger Bowl area north of Bozeman; he built log cabins for a living. In my experience, there are areas in Idaho that have a lot of Nazi/white supremacist/anti-government types. I don't like big government, but the Lord gave us human government for a reason. There are good people and bad people in government. I definitely don't like the Nazi/white supremacist mentality. Nazism is Satanic.

This guy dropped me off at the Sugar City exit and I found a camper near a construction site to sleep in. There were two or three blankets in the camper, so I was able to stay warm last night. My sleeping bag is good to around freezing, that is why when it is cold I am always looking for a haystack or a cornstalk stack or a vehicle or a building to sleep in--added protection from the bitter weather.

When Jesus was born over two thousand years ago, He was the greatest gift that God ever gave this broken, sin-sick world. There was no room at the inn, so Jesus was born in a manger in a pile of hay or straw. Wrapped in swaddling clothes. Lying in a manger because there was no room at the inn. No room at the inn. In the world system, the Kingdom of Heaven has no room at the inn. Sometimes there is room in the back seat of a junked car. Junked cars can be beautiful.

Meeting a Former Editor from Warner Brothers or Things Happen for a Reason

It was probably the spring of 1997. I hitchhiked north on U.S. 395 from southern California and got dropped off in Bishop. Bishop is a very beautiful place.

The mountains to the east were dry and brown, the mountains to the west (Sierra Nevadas) were rugged and snow-covered. There are a lot of irrigated ranches in that valley. I walked through Bishop for a couple of miles and then stopped north of town on U.S. 395. I waited for a short while and this vehicle pulled over to pick me up.

The guy who gave me a ride was probably in his late fifties or early sixties. He told me that he was coming from a ranch that he owned in Mexico; he was heading to Mammoth Lakes where he owned a grocery store. I told him that I was hitchhiking around the country for a short while; I had just quit my job at Harold Pike Construction Company in Ames, Iowa (Pike Construction hired me ten times in four years, I was grateful that they let me work for them so many times).

"So what did you do before you bought your ranch?" I asked.

"I worked for Warner Brothers as an editor," he replied. "I worked at Warner Bothers for a number of years and got tired of being in the studio."

"So what films did you work on?" I asked.

"One film I worked on was *High Plains Drifter*," he said.

I looked at him and exclaimed, "No way! *High Plains Drifter*? That is one of my favorite westerns. You are not going to believe this, but in 1995 I had a short story published by *Ethos* magazine. The title of my short story is 'High Plains Drifter.'"

"Really?"

"Yeah."

At the time, I had a few copies of my short story in a folder in my backpack. I would pass out my story to people if they were interested in reading it.

"When you drop me off, I will give you a copy of my short story," I said.

"Sounds good."

We drove north on U.S. 395. At Lake Crowley he turned off the road and dropped me off at this intersection. I dug out my folder that was in my backpack and gave him a copy of "High Plains Drifter."

"Thanks," he said.

"Thanks for the ride."

He drove off and I started walking up U.S. 395. I walked for a short while. The sun was down and I needed to find a place to sleep. I jumped over this fence and walked out into this sagebrush maybe a quarter of a mile from Lake Crowley. I rolled out my sleeping bag and slept there. I think it got down in the upper 20s F that night.

About my meeting the guy who gave me a ride from Bishop to Lake Crowley: there are no accidents in the Kingdom of Heaven. Things happen for a reason.

The next day I hitchhiked north to Reno.

[The film *High Plains Drifter*, starring Clint Eastwood, was made at Mono Lake near Lee Vining, California in 1973. Lee Vining is on U.S. 395 between June Lake and Bridgeport.]

A Conversation with a Vietnam Veteran

Back in November 2001 through August 2002, I hitchhiked in and out of St. John, Kansas quite a bit. St. John was my home base during that time. I would stay at one of a few places, do odd jobs and then I would hit the road.

A couple of people that I would stay with were a man and his wife. He was in his late fifties and she was in her early sixties. I don't remember their names, but let's call him Frank.

Frank was a Vietnam Vet who served in the U.S. Army in 1965-1966. He was exposed to Agent Orange and was on full medical disability. Frank was on his second marriage.

One day Frank and I were in the kitchen--I was sitting at the table and he was standing at the counter. I told him some of the things that I had experienced in my past: I went through a lot of rejection from family, friends and church people because of my Christian faith. Dad put me in mental hospitals, had me pay $5000.00 worth in hospital bills and then later told me that he paid for everything. My dad had absolutely no integrity whatsoever.

Frank then turned around and stared at me. He said, rather forcefully, "You're suffering from PTSD (Post Traumatic Stress Disorder)!"

I replied, "No way! You're crazy! I never was in the military and I never was in combat!"

Frank said, "You don't have to be in combat to have PTSD."

I said something like, "How can I have PTSD? There is no way I have PTSD." I was dumbfounded.

Then Frank got really angry and said, "I was in Vietnam. I saw many guys who were in serious firefights and you have the same symptoms as they do."

I didn't know what to think. The Lord puts people in your path for a reason. Maybe I was meant to hear what he had to say.

Eventually, I quit hitchhiking through St. John, Kansas and started hitchhiking in Wyoming, Idaho and Montana more often.

In the spring and summer of 2008, I passed through St. John, Kansas and tried to look up the people that I knew back in 2002; most of them had moved away.

I sometimes think back on that conversation. There may have been some truth to what Frank had said. I do know that through Jesus is great redemption. Repentance from sin and forgiveness for other people's trespasses are very powerful.

The moral of the story:

Don't call a Vietnam Vet crazy and. . .

. . . Sometimes a blind man doesn't know he is blind until someone tells him that he is blind.

My First Time in Jail for Hitchhiking

This is an account of my being arrested and put in jail for a short time in Riverton, Wyoming. I had failed to pay a fine ($60.00) or appear in court for a hitchhiking violation.

A couple of days ago, I hitchhiked from Jackson to Riverton, Wyoming. My last ride to Riverton was with this lady who sold Avon products. We had a nice talk. She said that she wanted to take me out to eat, so we had a buffet at a Chinese restaurant in Riverton. After the restaurant, she drove me to the south side of town near an industrial area. I thanked her and took my backpack and walked down this foot path. After about two hundred yards, I veered off the foot path and walked across this open ground to this place overlooking the Wind River. I set up my tent and bedded down for the night (or so I thought). It was around 9 PM.

A short while later, I heard this car driving around maybe a hundred yards north of my tent. I looked out of my tent and saw this car drive very fast in and then out of this gravel driveway.

Maybe fifteen minutes later I heard this other car drive down the same road. I looked out and saw the car turn and shine its headlights on my tent. The car approached my campsite and I got out of the tent to see what was going on.

The car stopped and a man and woman in uniform walked towards me. They were with the Riverton Police. They told me they were looking for some kids that were trying to break into a car in a housing subdivision just north of where I was camped. They thought that the kids may have been from the reservation (Wind River Reservation--made up of Arapahoe and Shoshoni Tribes) just across the river.

They asked me what I was doing and I told them that I was hitchhiking and had camped out for the night. They told me that there are a lot of violent crimes on the reservation; there were twenty-eight murders so far this year--probably alcohol and meth-related. I told them that I had camped here earlier this summer and that I was planning on hitting the road the next morning.

The lady police officer asked what my name was and she ran a check on me through the police department; I also gave them my driver's license. We talked for a little while longer and then she said that I had a Bench Warrant for my arrest and that I needed to pay sixty dollars or else go to jail. It was from a hitchhiking ticket I got back in February of 2009: I had failed to pay the fine or appear in court.

I walked back to my tent and looked in my billfold and told them that I had fifty bucks. She said that they needed sixty. Looked like I was going to jail.

They let me put on my pants and shoes and I took a few valuables with me. I was patted down for any weapons. They had me stand with my hands behind my back as they put these handcuffs on my wrists. They led me back to the police car and had me sit in the back seat. The handcuffs were very tight and uncomfortable.

On the way to the police station, they asked me if I knew anyone in Riverton that I could contact to help pay the remaining ten bucks. I gave them a name of a friend who had picked me up hitchhiking a couple of months ago.

We pulled into the garage at the police station. They led me to a large room with a table and sink. They had me empty my pockets and take off my shoes and sweatshirt. Then they led me into this small adjoining room and locked the door behind me. This room had a concrete bench to sit or sleep on; it had a sink and a toilet. It was probably ten foot by ten foot. I sat there for at least half an hour.

Then someone unlocked the steel door and they told me to come out. They said that my friend had arrived with the ten bucks. I put my sweatshirt and shoes back on and walked to this other room where my friend and his son were waiting. I paid my fine and walked out a free man.

My friend drove me to my campsite where I broke down my tent and put all of my gear in his pickup. We drove to his house and he let me sleep in his camper that night. I was very grateful that he helped me out with the ten bucks and for a place to stay for the night.

The police were friendly, courteous and professional; I was in jail for a very short time; I have no complaints there. I asked the police how long it had been illegal to hitchhike in Wyoming; they didn't know. I would really like to know WHY it is illegal to hitchhike in Wyoming.

****

Copy of Bench Warrant (Filed Mar 19 2009):
In The Circuit Court of The Ninth Judicial District
Fremont County, Wyoming
State of Wyoming, Plaintiff
VS.
Timothy M. Shey, Defendant

TO: ANY PEACE OFFICER IN THE STATE OF WYOMING: GREETINGS:

WHEREAS THE DEFENDANT, has done the following according to the Court record, more specifically set forth as follows:

Failure to Appear as ordered on 2/18/09

YOU ARE HEREBY COMMANDED to arrest the above-named defendant and bring him/her forthwith before this court to be dealt with according to law.

Bond: $60.00 [ ] Cash-- Must be posted before release from Custody.

[ ] Bond may be forfeited in lieu of appearance. The defendant may appear before this Court at 1:30 pm on Wed following his or her release.

Dated 3/18/09

Original Violation(s): 1)31-5-606 a SOLICIT ON STREETS & HWYS

On a Ranch Near Ennis, Montana

This past week I was hitchhiking in Montana and I ended up in Ennis. I went to the library and typed up some stuff on my *Digihitch* blog and then I walked to the Exxon gas station.

I was inside the convenience store buying something to eat, when this older man walked up to me and asked, "Are you the traveler? Is that your backpack out front?"

I said, "Yeah."

His name was Arthur and he said that he had done some hitchhiking in his younger days. He was originally from San Diego and did a lot of surfing at one time. Arthur used to hitchhike with a guitar. He asked me if I needed a place to stay for a while. He told me he needed some work done on his ranch and that he had a bad back; he had been in a real serious car crash years ago.

So I told him that that would be great and that I would like to work for him. I grabbed my backpack and we drove around six miles to his ranch. He had a housemate named Hal who had lived there for five years; Hal was married and divorced and pretty much retired. Arthur used to be a miner years ago.

I fed the horses hay and grain while I was there. Arthur and I hauled some garbage to the local dump and we did a lot of cleaning up of some trash in the house and rearranging some boxes for storage.

I ended up staying two nights and then hit the road. I hitchhiked south and made it to Driggs, Idaho where I met up with a friend. I stayed at he and his wife's place in Drummond last night.

Yesterday, I checked my email and Arthur sent me a very kind and thoughtful note; here it is below:

"Hello Saw man we are glad in the lord and holy power for leading you to us. We are very much lovers of good men who follow the path in life that few dare to seek, I find in you the good warm energy that god has bestowed upon you, follow your path no one else can, and remember us in your prayers we shall forever be in your kindness and have no regrets for the time you and we shared with you. Be always welcome in our tee pee. We enjoyed you and the god & man energy to shared with us. Have a safe and full filled life and some day return to us that we may share what god has given us to share with his chosen few. you are special in our hearts and minds so be good to yourself and we will not judge you but find in you faith to carry on and struggle with our human condition and remain thankfull to god first and the life of mammon second.

"your friends Arthur And Harold.

ps glad you liked my cooking. pax goldbear"

"No man has hired us
With pocketed hands
And lowered faces
We stand about in open places
And shiver in unlit rooms.
Only the wind moves
Over empty fields, untilled
Where the plough rests at an angle
To the furrow. In this land
There shall be one cigarette to two men,
To two women one half pint of bitter
Ale. In this land
No man has hired us.
Our life is unwelcome, our death

Unmentioned in 'The Times.'"

"When the Stranger says: 'What is the meaning of this city?
Do you huddle close together because you love each other?'
What will you answer? 'We all dwell together
To make money from each other'? or 'This is a community'?
And the Stranger will depart and return to the desert.
O my soul, be prepared for the coming of the Stranger,
Be prepared for him who knows how to ask questions."
"Remember the faith that took men from home
At the call of a wandering preacher."

--T.S. Eliot
Choruses from 'The Rock'

High Plains Drifter

By Timothy Michael Shey

[Fiction]

The big Kenworth roared west through Wyoming.

"So how long've ya been on the road?" the truck driver asked.

"A day or two," the young man replied.

"Where'd ya start out?"

"Western Nebraska. I was working on a ranch for a couple of days and got sick of it. I have a friend in California I want to see."

"California?"

"Yeah."

The truck driver was heavy-set and wore a short-cropped beard and baseball cap. The young man was slender and wore glasses. His only possessions: a backpack and sleeping bag.

"Ya got a long ways to go," the truck driver said. "I'll get ya to Salt Lake. Then I'm headin' north."

"Thanks for picking me up. It was cold standing out there."

"No problem."

The rugged, rolling terrain of Wyoming. The sagebrush. The dry air.

"So what'd ya do before the ranch?" the truck driver asked.

"I was in school in Manhattan."

"New York?"

"No. Kansas."

"Where ya from?"

"Garden City."

"I see."

The young man looked over the horizon to his right. There was silence for ten minutes except for the noise of the engine and the bounce of the tractor-trailer.

"So who's this friend of yours in California?" the truck driver asked.

"She's a poet."

"She?" The truck driver smiled and looked at the young man.

"I've never met her before. I've read a couple of her books and we've exchanged a few letters, that's all."

"I see."

"She has a daughter going to school in Santa Cruz that I thought I might like to visit, also."

"I don't know much about poetry. Is it like drivin' a truck?" the truck driver asked jokingly.

"Exactly." Exactly. Poetry is breath and fire and pain. Poetry is getting drunk or stacking hay on a ranch in western Nebraska. It is holding a beautiful woman in your arms; it is holding a baby in your lap. It is dropping out of high school because of the shallowness and stupidity. Exactly. Poetry is hitchhiking all the way to California to see a brilliant woman who loves the letters you write.

"So where'd ya stay last night? It got pretty cold out there."

"A rancher picked me outside of Laramie. He drove me to Rock Springs where his parents live. They let me stay overnight. Wonderful people. Gave me supper and breakfast."

"No kiddin'?"

"It was pretty incredible."

"I'll say. All a person hears about are people gettin' robbed or killed on the road."

"Yeah. Really."

The big Kenworth was going 80 miles per hour, passing cars and trucks. The speed and the power, the stress of steel and bolt, piston and axle and 18 wheels. Going west. Going west.

"So where you going after Salt Lake City?" the young man asked.

"Headin' north of Pocatello. Then I'll head back to Denver with another load."

Fire and breath and pain and heading north to Pocatello. Pocatello of your dreams. Pocatello of your nightmares. Six men die in gun battle with federal marshals at the Pocatello Corral. Southern Idaho desert. Dry heat, dry grass, dry blood on dry earth. Exactly. The breath of the moment, the heat of the battle--firefight in the Pocatello Corral.

One federal marshal wounded. Dry sun on another horizon. This is not Kansas. This is not Nebraska. This is Pocatello. Pocatello of your nightmares.

"This sure is wide open country," the young man said.

"It's a wasteland. Desert."

"I like wide open spaces."

"Then ya won't like California. Ever been to L.A. or Frisco?" the truck driver asked.

"No."

"Where does your poet friend live?"

"Big Sur."

"Never been there."

California of your nightmares. Big Sur of your dreams. Fire out of Kansas. Wheatfields and golden landscapes and dry air and blue sky and. Words, ink on paper, meter and fire. The anvil and the hammer and the fireblood of a wounded heart. Laceration and pain. Fire. The wordsmith labors and sweats and bleeds and brings forth new life. Anvil and hammer. The hot steel is shaped. Blow after blow. Sparks fly in the hot and dry air of Kansas.

"So how old are ya?" the truck driver asked.

"Twenty-three."

"So what do ya want to do with your life?"

"I want to be a bounty hunter or President of the United States."

The truck driver smiled and chuckled. "Sounds good to me. Ever see *High Plains Drifter* with Clint Eastwood?"

"I am the High Plains Drifter."

Flame out of Kansas. Riding west to the gold rush of your dreams. Desperate, unshaven, sunburned and hungry. Big Sur on your mind. Leather boots, leather skin, the stink of horse sweat. Shot six men in Pocatello just to watch them die. The bullet wounds of your heart, the anguish of the moment. Six men in Pocatello. Just to watch them die. You cinch the saddle down tight and ride west with the hot wind of Idaho at your back. You will ride west where the Pacific meets the edge of the Universe. There you will grow new muscle and ride a horse of a different color.

West. Flame out of Kansas. Exactly.
The big Kenworth rolled west through Wyoming and eternity.

Ethos
May 1995
Iowa State University

A House or a Home?

I was hitchhiking somewhere in southeastern Kansas a few years ago when I looked to my right and saw a big house out in the country. I told the driver of the car, "That's a nice house."

He said, "Yeah. Some people live in houses and some people live in homes."

It seems like most parents put their kids in public, private or parochial schools so that the teachers can baby-sit their kids. Parents are the primary teachers and caregivers. Parenting is a full-time job. Why would I want someone else to raise my kids?

Some people live in houses and some people live in homes. Some people spend more time taking care of their houses than they do with their wives or children. We are here to maintain and strengthen our relationships with God, our family and our friends than with our houses, our cars, our bank accounts, etc. If your life is cluttered with many things, then you will have to spend time maintaining these things. If you keep your life simple and uncluttered from things, you can spend your time maintaining your marriage and your family.

I was hitchhiking in Missouri a few years ago and this guy wanted me to meet his pastor. All that pastor could talk about was his new Volvo car. I wanted to ask him what the Lord was doing in his life, but since his relationship with God wasn't as important as his relationship with his Volvo, we talked about absolutely nothing.

Your treasure is where your heart is. You are what you eat; you are what you read. If all you think about is making money, then eventually you will make money. If all you think about is having a real nice car, eventually you will have a real nice car. If you continually chase God, then God will fill your cup to overflowing. Draw nigh to me and I will draw nigh to you.

I have been in some very nice, expensive houses throughout the United States, but most of the time, there was no peace in these houses. I have been in old, broken down cars and have had excellent Christian fellowship. When you die, you can't take your house or your car with you, but you will take your character with you. Either you have integrity or you don't. Integrity is your signature; integrity is who you are. Is your life gold, silver and precious stones or is it hay, wood and stubble?

You can tell a man by the house he lives in. Is it clean? Is it a mess? What books are in his library? Does he have a library? Is his house filled with beer cans and whisky bottles? Is he in debt up to his ears so that he can live in an expensive house? Is his house paid off? If a man cannot govern his own house, then why should he govern a city, a state or a nation?

I have been in some warm, wonderful homes in my journeys. These homes are built on a Rock and that Rock is Christ. The children are submitted to their parents, the wives are submitted to their husbands and the husbands are in submission to our Heavenly Father. These homes are peaceful and secure and a blessing to others. A Christian home is a powerful sermon and a powerful ministry.

Even though I do not own a house, I really don't feel homeless. I abide in Christ, so I can feel at home sleeping out in a pasture, under a bridge, in an abandoned vehicle or a haystack. Home is an attitude, a state of mind. Houses are for robots; homes are for Sons of the Most High God.

I am a Son of God.

Sans teeth, sans taste, sans house, sans everything.

Las Vegas Earthquake

Yesterday I hitchhiked from Elko, Nevada through Salt Lake City to Evanston, Wyoming on I-80. The two guys that took me to Evanston from Coalville, Utah bought me a meal at the Flying-J Truck Stop. From there I walked into town and checked my e-mail at the library. I then walked three miles north of town and found a pickup camper to sleep in. I believe it got down to 0 degrees F last night. It was also very windy; I am sure the wind chill factor was around -20 degrees F. Some snow even blowed into the camper. I stayed warm. (I think it is kind of humorous how the Lord always finds places for me to sleep.)

Last night, as I slept in the camper, I had a very vivid dream concerning the destruction of Las Vegas. I am guessing that an hour or so before this dream, I was attacked by Satan as I was sleeping. I was dreaming and in the dream I was walking down this street when all of a sudden I was tackled and thrown to the ground by a powerful, unseen force. This force or evil presence threw me down and beat me up pretty good—it even tried to pull the hair off of my head. When I woke up, I was still paralyzed by this evil presence and it was beating me up—the pain was real. The evil presence finally left and I went back to sleep.

Later on that night I had another dream and in this dream I was in a city square with hills all around it. It looked like a European city during the Middle Ages. There were many people in the city square. I was walking around with several friends. Then I saw the Pope (he looked like Pope Benedict—the present Pope) dressed in a white robe and wearing a white cap. My friends and I walked up to the Pope and I told him that Las Vegas was going to be destroyed. Then the Pope got this surprised look on his face and started to point at my friends and me and said, "I haven't seen you in church" (which meant they didn't go to a Roman Catholic Church). I then rebuked the Pope and told him that these people go to **Christian** churches. The Pope then got all bent out of shape and walked away.

I then separated myself from my Christian friends and walked through the town square. Then this woman yelled at me, "Shut up, Tim!" I walked past these Swiss Guards (that are used at the Vatican in Rome) (maybe I was in Rome) who were wearing funny-looking, striped uniforms and armed with swords. They were doing some outlandish rituals and I asked, "Why are you doing such ridiculous rituals?" and "Why are you doing such stupid bullshit?"

I walked away from the city square and the crowds of people and walked to the top of this hill overlooking the city. I looked in the direction of Las Vegas (it was on the side of the hill just opposite of where I was at) and all of a sudden the ground began to rumble and shake. I looked at Las Vegas as the earthquake swallowed it up. The cities and towns next to Las Vegas weren't even touched. When Las Vegas was destroyed, it looked like a piece of a jigsaw puzzle was removed and shoved underneath the rest of the puzzle. Then these huge rocks fell from heaven and pummeled the ground where Las Vegas used to be. I then raised my right arm in triumph and shouted, "Praise the Lord! Las Vegas in destroyed! Las Vegas is destroyed! Thank you, Lord! Las Vegas is destroyed! Las Vegas is destroyed!" Then the dream ended.

****

I left Evanston, Wyoming this morning and got back to Jackson around 4 P.M. On the way, I got dropped off just a mile or two from Alpine Junction and I looked up to my left and on this mountainside were many mountain goats. I then walked on down the highway and saw these two mountain goats peering over this jagged piece of rock that was probably a thousand feet above where I was walking. It reminded me of the phrase, "Hinds' feet on high places." I believe the Lord wanted me to see those mountain goats for a reason.

There are three things I have observed about mountain goats: they have no fear of heights, they have no real fear of heights and they absolutely positively have no fear of heights. They had such thick, white, wooly coats. The mountain goats on that mountain were such a beautiful sight; it was on the road that ran between Alpine and Hoback Junction on U.S. 26.

****

About being attacked by Satan last night: in my experience, sometimes Satan does a preemptive strike if he knows that the Lord is going to do something powerful in someone's life. Satan may attack you if the Lord is going to reveal something to you, or if the Lord is going to deliver you from demonic bondage or if the Lord is going to heal you, etc.

I believe that the Lord is going to destroy Las Vegas soon—maybe in December of 2006.

A Fast Trip

I just got back from a hitchhiking trip that went very fast. I left Jackson yesterday and got a good ride to Shoshoni. This guy's name was Richards and he used to play college and pro football in the sixties and seventies. He was an attorney in Hawaii and had retired to Jackson, but his wife doesn't like the cold weather so they may go back to Hawaii. I told him that I took my LSAT back in 1994, but did poorly. I applied to three law schools, but did not get in. It didn't break my heart. I had enough of the liberal dominated academic establishment and I'm sure that the United States has plenty of lawyers, so I guess I'll be that living sacrifice and not go to law school. I did like the Drake University Law School in Des Moines; it had a brand new law library. I thought maybe I would like to study constitutional law, but as we can see, the Lord had different plans.

I then got a ride to Cody from a guy from Nebraska. From Cody I got a few rides to I-90 in Montana. And my last ride that evening was from a Christian who knew a couple of people that I knew in Absarokee.

I slept in a cab of a truck last night in Columbus. Then I walked at least a few miles west on I-90 and this Christian driving a tractor-trailer fuel truck picked me up and took me all the way to Livingston where I checked my mail. There was no mail (I don't get much mail anyway) and the Lord told me to change my mailing address to Jackson.

I got a ride from Livingston to Bozeman and then to Big Sky. From Big Sky I got a ride to Sugar City, Idaho. It took me two rides to get to Tetonia when this young lady named April picked me up. She told me that she had broken her arm a while back—a compound fracture—and that the rehabilitation had been painful. I told her that I would like to pray for her so that her arm would heal up good. We drove to Driggs and parked at the grocery store parking lot and I prayed in the Holy Ghost so that her arm would heal up faster. She was grateful.

I got a short ride to Victor, and at Victor, I got a ride with a young man from British Columbia, Canada all the way to Jackson. We had a great talk. He graduated from Denison College in Ohio with a degree in geology. I would love to see British Columbia someday. I hear it is very beautiful.

Why the fast trip? The Lord works in strange ways. It somewhat confounds me why the Lord wants me to change my mailing address so much. What is the reason? What is the significance? Does the Lord do it just so that He can see that I am obeying Him? Well, anyway, I did meet some interesting people and I was able to pass out my book and journal (which is copied on a mini CD) to five or six people.

I was walking around Columbus, Montana last night looking for a place to sleep. I walked over to this building with stacks of lumber in it. I have slept there before, but there just wasn't any place to sleep (they had rearranged the lifts of lumber from the last time I had slept there). So I walked on past the Timberweld place and found a truck—a dump truck—and slept there last night. It rained off and on through the night. It was cozy sleeping in the cab of the truck.

I had a real good talk with the Christian who drove the tractor-trailer. He told me about some of the dreams that the Lord had given him. In one dream he saw some nuclear explosions going off somewhere. He told me that he had lived a totally different life, until the Lord had got a hold of him. The Lord really stripped him of everything he had and of everything that he thought was important to him (which is what the Lord did with me) and began to rebuild his life.

It was such a fast, crazy trip. Sometimes it takes forever to get a ride. Whether I get some place fast or get some place slow—it is always perfect timing with the Lord.

A Dog Named Patton

Yesterday, I hitchhiked from Riverton through Shoshoni and Thermopolis to Cody. From Cody I walked a few miles and got a ride with a truck driver named Steve. Steve was from North Dakota and he had spent eight years in the National Guard and had spent time in Kosovo and Iraq as a combat engineer. His job in Iraq was to find roadside bombs and get rid of them. He had been blown up four times in his Humvee. One time his Humvee stopped right next to a roadside bomb and it did not go off. Right then, he told me, he felt invincible—he wasn't meant to die there and it made him a believer in God. Steve then told me what his grandfather had told him: "If you are meant to hang, you will never die in a fire." Which means: God is sovereign.

Steve also told me that he was raised in the Catholic Church, and before they got confirmed, they had to go to this class—I guess, to explain Catholic doctrine. In the class, Steve told the priest, "If I am sitting in a goose blind thinking about God, isn't that better than sitting in church thinking about being in a goose blind?" The priest kicked him out of the class.

Steve had his pet dog—it looked like a black lab cross—in the cab with him. He named his dog, Patton, after General George S. Patton of World War II fame. We talked a lot about the war in Iraq.

Steve took me from Cody through Big Timber, Montana and drove north on U.S. 191 to Harlowton. From Harlowton we went west on U.S. 12 where he dropped me off at the junction just north of Two Dot, Montana. It was after sundown, so I walked two or three miles till I found a haystack. I slept in the haystack last night. I believe it got down to 12 degrees F. It was a beautiful, crisp, cold night; there wasn't a cloud in the sky; the stars were very bright. I was grateful for the haystack.

This morning I walked maybe two miles west on U.S. 12 and got a ride with two women in a van. They lived in Harlowton and were going to do some skiing north of White Sulphur Springs. They were

Christians and we had a pleasant conversation. We talked about hitchhiking and living by faith. They dropped me off here in White Sulphur Springs.

Some Monks Hitchhike

Some inspiring words from Fyodor Dostoyevsky.

Excerpts from *The Brothers Karamazov* by Fyodor Dostoyevsky; Book VI, "The Russian Monk":

"Fathers and teachers, what is the monk? In the cultivated world the word is nowadays pronounced by some people with a jeer, and by others it is used as a term of abuse, and this contempt for the monk is growing. It is true, alas, it is true, that there are many sluggards, gluttons, profligates, and insolent beggars among monks. Educated people point to these: 'You are idlers, useless members of society, you live on the labour of others, you are shameless beggars.' And yet how many meek and humble monks there are, yearning for solitude and fervent prayer in peace! These are less noticed, or passed over in silence. And how surprised men would be if I were to say that from these meek monks, who yearn for solitary prayer, the salvation of Russia will come perhaps once more! For they are in truth made ready in peace and quiet 'for the day and the hour, the month and the year.' Meanwhile, in their solitude, they keep the image of Christ fair and undefiled, in the purity of God's truth, from the times of the Fathers of old, the Apostles and the martyrs. And when the time comes they will show it to the tottering creeds of the world. That is a great thought. That star will rise out of the East."

"The monastic way is very different. Obedience, fasting, and prayer are laughed at, yet only through them lies the way to real, true freedom. I cut off my superfluous and unnecessary desires, I subdue my proud and wanton will and chastise it with obedience, and with God's help I attain freedom of spirit and with it spiritual joy."

San Miguel, California

I got dropped off in San Miguel, California late yesterday afternoon. I slept in a haystack on a ranch just north of San Miguel. It got down to 16 degrees F last night--very cold for this part of California.

I left Riverton, Wyoming on the 10th of January. I hitchhiked south through Utah and Arizona. I stayed at a Christian mission in Flagstaff for one night. I headed west on I-40 and got a ride from Ash Fork, Arizona to Kramer Junction, California with a couple, Stephan and Suzie.

Stephan, Suzie and I had a great conversation. They were coming back from Phoenix and Flagstaff and were going back to Orange County where they lived. Stephan used to do some hitchhiking years ago. We talked about living by faith and other things of God. They bought me a big Subway sandwich in Kingman and then they had me drive from Kingman to Barstow.

I slept in some abandoned motel in Kramer Junction and hitchhiked to Bakersfield and then to Wasco. A young man named Steve gave me a ride to Wasco. We had a good talk about the Gospel. He said that you had to be desperate for God when you live as a Christian. I like the word “desperate”; if Steve stays hungry and desperate for God, then he will always be in good hands--in God’s hands. I slept in an orchard just west of Wasco that night. The next day I made my way to Lost Hills, Paso Robles and then to San Miguel. I will continue to go north on U.S. 101 till the Lord tells me different. It is still very cool here in San Miguel; I hope it warms up some.

Acts 2: 38

Just got dropped off here in Buckhorn, California. I am on Highway 88 going east towards Nevada.

When I left San Miguel, I got a ride to King City where I slept in a junked pickup that night. The next day I got a ride with a guy named Tom and we went to his office in Salinas where I put my files on his computer and he made me a sandwich. Tom took me to Gilroy and dropped me off at the library so that I could send some e-mails.

I walked east out of Gilroy and got a ride with a Christian, Jose. He took me out to eat in Los Banos. Jose then took me home so that I could shower, shave and wash my clothes. His wife came home later that evening. I was so grateful to get a shower and sleep on the couch.

The next day I got a ride to Merced. From Merced I walked north to Atwater and slept near the railroad tracks. The following day I walked for a while and got a ride to Modesto on U.S. 99 with a Christian, Nathan, who gave me some money so that I could get something to eat. I slept in somebody's barn near Manteca that night. There are a lot of orchards in that part of California.

This morning I walked for a while and got a ride to Stockton and then to Jackson on Highway 88. I just got a ride from Jackson to Buckhorn with a couple of guys who wanted to read my *High Plains Drifter*, so I'll e-mail it to them when I run into a library.

Earlier today as I was riding from Manteca to Stockton on U.S. 99, I saw "Acts 2: 38" on a bumper sticker.

Acts 2: 38: "Then Peter said unto them, Repent, and be baptized every one of you in the name of Jesus Christ for the remission of sins, and ye shall receive the gift of the Holy Ghost."

A Foot Soldier

Last night I had a dream where I was a foot soldier in a large army--it was an army right out of the Middle Ages. I had a spear in my right hand and I was marching with thousands of soldiers. I believe there were also soldiers on horses. We were marching towards this higher elevation where this Muslim army was waiting for us. I remember I was a little nervous marching into battle. Then in faith I ran to the front of our army and began praying in tongues. The Muslim army began to throw spears and release their arrows and I would hack at the spears and arrows with my sword. After the Muslim army's first volley of artillery, I wasn't nervous anymore. Our army ran up the hill as I was praying in tongues. That is all I remember about the dream.

A Hitchhiker in Bakersfield

Just got dropped off here in Wells, Nevada. I got a good ride with a truck driver from Sparks--his name was Viktor and he was originally from the Ukraine. He didn't speak much English, so we didn't talk about much.

Yesterday as I walked through Mariposa, California, these two ladies picked me up and dropped me off at Mt. Bullion--there was a bar there, so I thought I would stop there and get a cheeseburger.

These two ladies were Christians and the lady driving knew that she was supposed to give me a ride. She told me that she picked up this hitchhiker in Bakersfield a while back and he was really different--he kept staring at her. He asked her, "Aren't you afraid of me?" And she said, "I am washed in the Blood of Jesus Christ." He didn't say much after that.

The lady then told me that later on--it was either in a newspaper article or on the local nightly news--that the police had a man in jail that had killed a few women in the Bakersfield area. The hitchhiker that she had picked up looked exactly like the guy that the police had in custody.

The police questioned the hitchhiker/killer and he said that he took this one woman out to Tehachapi into the desert and he wanted to kill her, but he wasn't able to. My guess is that she was a Christian and the demon inside of him was not able to overpower the Holy Spirit inside of her. There is power, power, wonderworking power in the Blood of the Lamb.

Some hitchhikers really make it difficult for other hitchhikers, but the Lord is with me--He inspires people to give me rides when I need them. I am a Blood-washed hitchhiker.

Lucille

I hitchhiked from North Fork, Idaho to Missoula, Montana today. It is very hot right now--around 96 degrees F--so I thought I would sit down at a picnic table under a shade tree and do some writing.

Yesterday I was walking somewhere between Yankee Fork and Clayton, Idaho on U.S. 93 when this lady pulled over to pick me up. Her name was Lucille and she was 81 years old--and she was an on-fire, Holy Ghost Christian. We had a great talk all the way to Salmon where she stopped to visit a friend of hers, Dorothy, who is also a Christian.

Dorothy had been suffering from macular degeneration in her eyes, her ears weren't 100 per cent and she had arthritis in her hands. Other than that, Dorothy looked pretty healthy for a woman of 84. She definitely was anointed with the Holy Ghost and had a very strong faith in the Lord.

Lucille and I had Dorothy sit in a chair in the living room and we began to pray for her. I laid my hands on her head and began praying loudly in tongues. I then laid my hands on her ears. I could feel virtue go through my hands. Dorothy definitely had a healing touch from the Lord. It was a powerful time.

After the prayer meeting, Dorothy tried to stand up, but she wobbled around a bit. I asked her if she was dizzy. Dorothy said, no, that she was drunk in the Spirit.

Dorothy and Lucille took me out to eat at Brewster's Restaurant in Salmon and then I hit the road. I hitchhiked to North Fork and then walked a few miles, jumped over a fence and slept in someone's pasture that night. I found an old piece of plywood lying around in the grass, unrolled my sleeping bag on top of it and slept pretty good last night.

I am at the Missoula Memorial Rose Garden; they have a World War II, Korea and Vietnam War memorial here. I'll probably walk to I-90 and then head east. It is really hot here [it later got up to 102 degrees F]; I'm feeling a bit dizzy.

A little bit more about Lucille. She said that she was first married in 1942. Her first husband was in World War II and was a waist gunner in a B-17 bomber. He flew in 50 missions over Europe. On his 50th mission, he got shot down over France and was captured by German SS troops and spent two years as a P.O.W. He said the flak was really bad: the tail section of the B-17 got blown off; half the crew was killed. The pilot, the co-pilot, another guy and he were able to bail out of the plane. The French Underground tried to help them, but they were eventually captured. One guy did manage to escape and made it back to England and then to the United States.

I guess Lucille's husband was tortured while he was a POW. When he came home after the war, he suffered greatly from battle fatigue: she would be sleeping in bed and he would start hitting her because he would be having a bad dream or a flashback. After ten years of physical abuse, Lucille got divorced from her husband. In 1945, World War II ended--but not for some people.

Lucille later got remarried and got gloriously saved at the age of 47. Now she joyfully serves the Lord and does His will. She even picked up a hitchhiker on a highway in Idaho and I am very grateful that she did. Lucille is a gift from God.

A Speed Skating Coach, a Dream and a Former Drug Dealer

Just got back from a fast trip. I hitchhiked out of Jackson on the 19th of July. I was walking north of Ashton, Idaho when this tractor-trailer pulled over to pick me up. I climbed up into the cab and the truck driver said that he had picked me up before. His name was Stan and after a few minutes I recognized him. He said he picked me up in Billings, Montana a year or two ago and took me to Sheridan, Wyoming. Stan said that he never picks up hitchhikers, but that he has picked me up twice now.

I remember the first time Stan picked me up he told me that he was a speed skating coach in Holland. He was born in Holland and then was raised in Poland. His English wasn't so good, so it was a bit difficult to understand him. Stan said that he coached speed skating in Holland for fifteen years and then came to America and coached speed skaters here. He helped coach Ard Schenck (Olympic Gold Medalist from Holland), and later, Eric Heiden and Dan Jansen (Olympic Gold Medalists from the United States). He is now retired from coaching and drives a truck for a living. Stan lives in the Salt Lake City area. He dropped me off in Bozeman, Montana and I slept in that junked van for the night.

The next day I walked outside of Bozeman for a mile or two and got a ride with a truck driver all the way to Norfolk, Nebraska. I then hitchhiked south through Columbus and made it to Stromsburg where I slept at a construction jobsite a few miles south of town. That night I had a dream. In the dream I was in a room (it might have been a bathroom) in a house that was filled up to the shins with water. There were a lot of clothes floating in the water--looked like someone's dirty laundry. I saw a dark cloud in the water. A former housemate of mine from Ames, Iowa was in that same room. We lived together in the same household with a few other Christian men for two and a half years. I walked up to him and asked him, "Have you repented of your sin?" He didn't say anything. Throughout the dream he was always looking down--he never looked up at me--it looked like he was convicted of sin. I later found the drain, uncovered it and all the water drained out of the room. I walked out of the house and then the dream ended.

I was a bit curious as to why the Lord would give me a dream about a former housemate whom I have not seen since the early 1990s--we lived in the same house from 1987 to 1990. It was just over twenty years ago when I had hitchhiked from Ellensburg, Washington to Ames, Iowa. I think I arrived in Ames around 10 July 1987; within a week I got my job back at Hanson Lumber Company. I lived in that same house for seven years. The two and a half years I spent with the above housemate were unpleasant: he was very carnal, he went to a church that hated the power of the Holy Ghost (praying in tongues, healing, deliverance from demons, etc.); we really had nothing in common; he made my life fairly miserable while I was living there. We were living in a Christian household, but I was always leaving the house looking for Christian fellowship--I would even go out into the woods and pray and praise the Lord--I didn't know what else to do. Why would the Lord give me a dream about someone I knew twenty years ago? I am sure that in time the Lord will reveal more to me about this dream.

The next morning I got rides south on U.S. 81 to Salina, Kansas. I waited for quite some time in Salina and finally got a ride to Colby, Kansas on I-70. I then got a ride to Seibert, Colorado where I got a cheeseburger. From Seibert I got a ride with a guy named Dave all the way to Jackson, Wyoming.

Why the fast trip? The Lord's thoughts are higher than our thoughts; the Lord's ways are higher than our ways. The trip from Montana to Nebraska went very fast and was very blessed--the truck driver even paid for my supper in Gillette, Wyoming and my breakfast in Murdo, South Dakota; in Belvidere, South Dakota, he stopped at a rest area and slept in his sleeper and I jumped over the fence and slept in a grassy field that night. After he dropped me off in Norfolk, I got a ride with two Hispanic teenagers to Columbus. The one guy told me that he was very interested in my travels and that he really wanted to learn to speak English well--he was going to be a sophomore in high school this coming fall--he had come from Mexico just three years ago. South of Columbus this man and woman and their three kids picked me up. He used to be a drug dealer and had spent some time in prison. He told me that he was a real bad ass at one time. He told me that he once drove to this guy's place so that he could kill him, but the other guy sprayed his car with bullets and two bullets grazed his arm and shoulder. He told me that he should have been dead. I told him that the Lord protected him because He had a purpose for his life. After he got out of prison (he became more committed to Christ in prison), he got a good job and now is taking care of his wife and children. I told him that the Lord was using him as a light for the Gospel in his hometown. We had a real good talk. Maybe the Lord had me hitchhike so quickly from Montana to Nebraska to talk to that Hispanic kid and that former drug dealer (I gave both of them my CD). If the Lord wants you to get someplace fast, you get there fast, let me tell you.

Looks like I will head up into Montana tomorrow.

Stacking Hay in Ashton, Idaho

For the past two nights I have been staying with Gene and Shauna and their three kids on a small farm near Ashton, Idaho. I met Gene last week while I was hitchhiking north of St. Anthony. Gene and I have hauled two loads of hay from Hamer to his farm. I haven't stacked hay in years and years. Growing up on the farm in Iowa, I baled a lot of hay between the ages of eleven and sixteen. This is really helping out Gene because he lost a leg in a car crash several years ago and he has a bad back. I was down to my last dollar when I got to his place, so a little extra money is going to help out a lot. We have one more small load to get this afternoon, so I will probably hit the road tomorrow and head north into Montana.

They Are Fighting People

On the 5th of October, I slept outside in my tent just a few miles west of Barstow, California. The next morning I walked several miles west on Highway 58 till this truck driver picked me up. His name was Chi and he was originally from Thailand. He had been in the United States since 2001. Chi drove to Bakersfield where he took me out to eat at a Vietnamese restaurant. We were driving north on Highway 99 when Chi asked me where my family originated. I told him that my ancestors came from Germany and Ireland. When Chi heard the word "Ireland," he said. "Oh, you are Irish. They are fighting people." I began to laugh and I told him that SOME Irish like to fight.

We drove through Fresno and then he stopped at a truck stop at Ripon, California. Chi got me a shower ticket, so I was able to get cleaned up. From there we parted and I walked north to the outskirts of Manteca where I slept in an orchard that night.

The next day I was walking on Highway 88 heading east towards Carson City, Nevada when this guy picked me up and took me to a small town called Lockeford, California. We stopped at a small shop that sold meat and sausage. As we waited in the shop, I noticed this street sign on the wall behind the cash register. I casually saw the word "Rush" on it, so I assumed it meant "Rush Hour" or something. On closer inspection, the sign read: "Reserved for Rush Limbaugh fans only." I thought it was pretty funny. This guy bought me a big piece of teriyaki beef jerky--it was the best beef jerky I have ever had--it lasted me for the next couple of days. That shop is supposed to have the best meat and sausage in the area.

After Lockeford, I got dropped off in Clements and walked a number of miles on Highway 88 to the Ione turn off. I finally got a ride to Pioneer where I walked to a grocery store and bought two packages of raisin bagels.

The next day I got a couple of rides to Lake Tahoe, Nevada. There I got a ride with a beautiful young lady named Dee. She drove me to Carson City where we stopped and I prayed for her. She was drinking alcohol and seemed pretty depressed and in pain, so I talked about the Gospel as much as I could. We stopped at this city park and sat down at this picnic table and talked about various things. Dee then dropped me off and I walked to the north side of town where I got a ride to Reno.

This guy's name was Jim. We were talking for several minutes and then he looked at me and exclaimed, "Hey, I picked you up earlier this summer!" Jim then asked, "Didn't you give me this disk with your writings on it?"

"Yup, that was me," I replied. We drove to Reno where he went to see his daughter. Jim's daughter was staying with this lady and her daughter. Jim's wife was addicted to crack cocaine, so his daughter was staying there. We stayed at their place for half an hour. The lady gave me a nice supper and then Jim and I left. Jim dropped me off at the final exit east out of Sparks. I slept at this construction bone yard that night.

Escape from Cuba

Just got dropped off in Bishop, California. Looks like there is a big snowstorm coming into the Sierra Nevadas. The storm may last for two days.

Yesterday, I got a ride with a guy named Robert. He took me from just west of Flagstaff, Arizona on I-40 to Barstow, California. He was originally from Havana, Cuba. Robert spoke in broken English. He escaped from Cuba three years ago. He and seven other people got in a boat and began rowing to Florida. They were in the boat for eight days. Five of the people died enroute. Robert, another man and a woman survived the trip. It took a lot of guts to do that. Freedom--or the hunger for freedom--is a very powerful mover in a world of slavery and oppression. The human spirit cannot remain shackled forever. Jesus sets the captives free.

Sitting in Jail in Broadus, Montana

I am now sitting in the jail at the Powder River County Sheriff's Office in Broadus, Montana. Looks like I will stay here overnight and then head west tomorrow. I hitchhiked from Rapid City, South Dakota to Broadus earlier today. It is snowing pretty good outside, so I have a warm place to sleep tonight. When I got dropped off here in Broadus, I went to the Sheriff's Office to see if someone could put me up in a motel for the night. The sheriff suggested that I stay in the jail because it was empty and it had just been cleaned. This will be the first time I have ever stayed in a jail.

Tim is the guy who gave me the ride from just west of Rapid City to Broadus. He is originally from Seattle where he used to work as a sheet rocker. He and his wife and three children live here in Broadus so that they can take care of Tim's grandfather.

Tim told me that he became a Christian a while back after he was in a really bad car crash. His buddy died in the crash and Tim was in a body cast for two months. Crisis sometimes precedes conversion.

We stopped in Belle Fourche, South Dakota where Tim bought me a hamburger at Hardee's. He told me that his grandfather was 88 years old and was a veteran of the Battle of the Bulge and the Battle of Huertgen Forest in World War II. I told Tim that his grandfather should write his experiences down before he passes away. Tim told me that his grandfather really didn't want to talk about the war. His grandfather used to crawl up close to the German lines and call in artillery on their positions. I'm sure he saw a lot of horrible carnage during the war.

So tomorrow I will either hitchhike to Miles City and head west on I-94 or else I will hitchhike through the Northern Cheyenne and Crow Indian Reservations and then take I-90 to Hardin and Billings. The snow storm is supposed to last into tomorrow morning.

Abraham from Macedonia

This morning I walked around five miles east on I-80 and finally got a ride to Fernley, Nevada with a Christian. I was broke, so he gave me a ton of change—nickels, dimes, quarters and pennies—and I got a couple of sandwiches and something to drink in Fernley. From there I walked two or three miles and got a ride with a truck driver named Abraham. We had a great talk. Abraham was originally from Macedonia. He has been in the United States for eight years. He asked me a lot of questions about my Christian faith. Abraham told me that he was a Muslim. We talked a lot about Abraham, Isaac and Ishmael.

I told him that it was important to know that Isaac was the son of promise (faith) and that Ishmael was the son of the flesh (non-faith). The Messianic Line came through Abraham, Isaac, Jacob and the Twelve Patriarchs. Abraham (the Macedonian) was very receptive as I spoke to him about having faith in Christ; how His blood cleanses us from sin and that the Kingdom of Heaven is a spiritual kingdom. I wanted to tell Abraham that he was not far from the Kingdom. It was a very edifying conversation and I am very grateful to the Lord that Abraham gave me a ride. He dropped me off here in Winnemucca and was going to take his tractor-trailer all the way to Lewiston, Idaho.

A Hitchhiker, a Knife and a Piece of Paper

Yesterday, I was walking on U.S. 83 just east of Rexford, Kansas and this young lady named Becca picked me up. She drove me to Oberlin. Becca and I had a real good talk about the Gospel; she was a believer.

Becca told me about this Christian man and woman who picked up a hitchhiker. They took the hitchhiker to some town, bought him supper and paid for his own motel room. The couple was also staying at the same motel as the hitchhiker.

The next morning, the man and woman walked to their car and on the front seat was a knife on top of a piece of paper. On the piece of paper, the hitchhiker left a message and he said that he was planning on killing them, but since they gave him supper and a motel room, he didn't want to kill them. The hitchhiker thanked them for their hospitality.

The hitchhiker giving his knife (a potential lethal weapon) to the man and woman was an act of repentance.

A Ride in Nebraska

Yesterday, I was somewhere between Bridgeport and Alliance, Nebraska when this car pulled over to pick me up. There were three guys in the car; I got in the back seat. As we started to drive down the road, the guy next to me asked, “Aren’t you from Ames, Iowa?”

I looked at him with a surprised look and said, “Yeah. How did you know that?”

He said, “I picked you up hitchhiking a few years ago and gave you a ride to Alliance. You made a photocopy of your book [*High Plains Drifter*] and gave it to me.”

I was stunned. We shook hands and then he said, “My name is Harold. I read your book and really enjoyed it. I passed it around to some friends of mine.”

It’s a small world. I remember making a photocopy of *High Plains Drifter* in Alliance for somebody, but I think it was more than a few years ago. I told Harold that he probably picked me up in 2001 or 2002.

So they drove me to Alliance and took me out to eat at a Mexican restaurant. The guy who was driving was Doug. Doug owned a junkyard nine miles from town; he let me and Harold stay at his place last night and Harold bought me breakfast this morning.

While we were eating breakfast, Harold told me that he was hitchhiking in Missouri back in the 1970s and William Least Heat-Moon picked him up and gave him a ride to Iowa. William Least Heat-Moon later wrote the book *Blue Highways*. I believe *Blue Highways* was a bestseller in the early 1980s. While I was living in Venice, California in the spring of 1984, I read *Blue Highways* and thought it was a very good book. I wrote William Least Heat-Moon a letter telling him how much I liked his book; he wrote me back, but I no longer have a copy of this letter.

Barack Obama and the Media

Last night I had a very strange, violent and disturbing dream about Barack Obama. In the dream, I was staying at this homeless shelter in Riverton, Wyoming (whenever I was hitchhiking through central Wyoming I would stay at that shelter from time to time). There were several people at the shelter. Then Barack Obama showed up—he was smiling, shaking hands with people and campaigning for the Presidency of the United States. It must have been nighttime because I laid out my sleeping bag on the floor to get some sleep.

Then there was this commotion in the other room. It sounded like there was a struggle between two people. I ran into the other room and Barack Obama had a rifle in his one hand and had grabbed Diane Sawyer (a well-known journalist) with his other arm. Diane Sawyer was struggling greatly and was trying to cry out, but Barack Obama had his hand over her mouth.

Then Barack Obama forced Diane Sawyer outside where they got into a car—it was a convertible. There was this homeless man nearby who looked like he wanted to help Diane Sawyer, but he was not able to because Barack Obama was holding a rifle. Then Barack Obama forced the homeless man into the car at gunpoint.

The last scene: I ran outside after Barack Obama, but he pointed a .44 magnum at me as he drove past in the convertible. I crouched behind another car as he drove past. Barack Obama was full of hate and anger and violence.

Paga

INTERCESSION (*Strong's* 6293 ***Paga***)

Paga: Hebrew for "intercession," has many different meanings which help us to understand intercession. Listed below are six different ways *paga* is translated which help in better understanding intercession.

1. ***Paga***: (Judges 8: 21; I Samuel 22: 17-18; II Samuel 1: 15; I Kings 2: 29)

In all these verses, the Hebrew word *paga* is translated "to fall upon" meaning to kill or destroy. These verses all refer to obedience to "fall upon" the King's enemies at the King's command.

So we are called to "fall upon" the King of King's enemies (which are demon powers) and *destroy their works.*

2. ***Paga:*** (Genesis 28: 11, 16; Job 36: 32)

In these verses, *paga* is translated to "light upon", meaning to hit the exact place God intended. The first example is Jacob, who just happened to "light upon" (*paga*), the exact place God wanted him to. After God had spoken to him, he confesses to the fact that God is in this place, and he didn't know it. God had caused him to "light upon" a certain place where Jacob could be spoken to.

The second example is in Job and should be read in many translations. The *New International Version* states "He fills His hands with lightning and commands it to *strike its mark.*"

The New International Version translates *paga* to "strike its mark." This means it hit exactly where God intended.

God-causes are *paga*, intercession, *to hit the exact place needed.* Like Jacob, we might not know we are in the exact place God wanted us to be in, we might have just prayed in a certain way or spoke in the Spirit. Then we find God has caused us to (*paga*) hit the exact mark. Compare this with New Testament verses: "Likewise the Spirit also helpeth our infirmities, for we know not what we should pray for as we ought, but the Spirit itself maketh intercession for us with groanings which cannot be uttered. And he that searcheth the heart knoweth what is the mind of the Spirit, because He maketh intercession for the saints according to the will of God." (Romans 8: 26-27)

3. ***Paga***: (Exodus 23: 4; Joshua 2: 16; I Samuel 10: 5)

In these verses, *paga* is translated *"to meet" as in contact.* The first time *paga* is translated "to meet" is when a lost animal is met, the finder should return it to its owner. We in intercession contact lost souls and pray them back to their Creator.

4. ***Paga***: (Joshua 19: 11, 22, 26, 27 & 34)

In these verses, *paga* is translated "reaches" referring to boundaries set up for each tribe of Israel. The land they were given reached from one point to another.

God-causes are *paga*, to "reach" all of the appointed blessings He has in store for us. When we are restricted from our God-given blessing (possessions), we should intercede (*paga*) and the intercession will deal with the restriction.

5. ***Paga***: (Judges 18: 25)

Translates to "run" upon and destroy. In this verse, you see the violent force of intercession (*paga*).

6. ***Paga***: (Isaiah 53: 12; 59: 16; Jeremiah 7: 16; 27: 18; 36: 25)

In these verses, the word *paga* is translated "intercession". God reveals in these verses what to pray (intercede) for and what not to pray for. (Jeremiah 17: 16)

Intercession is a combination of understanding prayers and spiritual praying or praying in the Spirit.

Conclusion: The word ***paga*** translates many ways and when taken together, a powerful type of intercession is seen.

1. An intercession that destroys the King's enemies.
2. An intercession that hits the exact mark.
3. An intercession that is involved with praying for the lost.
4. An intercession that sets boundaries.
5. An intercession that is violent against the kingdom of darkness.

[The preceding information regarding intercession was provided by Lou Somerlot. Lou picked me up hitchhiking south of Missoula, Montana.]

Sleeping on a Stack of Lumber in Columbus, Montana

Yesterday I hitchhiked from Belgrade to Columbus, Montana. I slept on a lift of lumber in a shed at the Timberweld place last night. The stack of lumber I slept on was three lifts high; the stack in front of me was four lifts high, so I was well-hidden from anybody at ground level.

This morning around eight o'clock I heard some people talking in the shed that I was in. There were two men and one woman. They were taking inventory. When they got near to where I was sleeping, this one guy climbed up a ladder to read off the numbers to the lady below.

Then the guy on the ladder saw my shaving kit and my shoes. "Hey, I see something," he said.

"What is it?" the other guy asked.

"I don't know. Let me get my flashlight." He shined the flashlight on my things and said, "I see a pair of shoes."

The other guy asked, "Do you see a blanket? We had a guy sleeping in here a while back. If you see anybody up there, apologize [for waking him up] and move on to the next stack."

That's when I spoke up. "Hey, I'm up here. I'll get out as soon as I can." I had been eating some bread and peanut butter for breakfast.

The guy on the ladder said, "That's all right."

The two men and the woman continued with their work.

After I packed up my things in my backpack, I jumped down from my sleeping berth and walked over to the man and woman doing inventory—the other guy had walked off to some other place. We spoke for a little while; the lady said that it had gotten down to 18 degrees F last night. I stayed nice and warm on the stack of lumber. I then walked across the railroad tracks to a convenience store to get a cup of coffee.

I will head south to Red Lodge later this morning and then mosey into Wyoming.

Men Plan and God Laughs

For the past few nights I have been staying at Kim and Pat's place in Stites, Idaho. Kim and I have been cutting up a number of logs on his portable band saw. We have been cutting up white pine, yellow pine and red fir and making it into boards for himself and for some of his customers.

Today I noticed in the *Sunday Lewiston Tribune* (18 April) these headlines on the front page: "Ash may hover for days over Europe"; "Volcanic eruption in Iceland continues to snarl world plans."

In today's *Lewiston Tribune*: "European airlines are becoming impatient"; "Hundreds of millions of dollars being lost each day as volcano in Iceland continues to disrupt air traffic."

The Lord really knows how to slow people down. God's thoughts are higher than our thoughts; our plans are not necessarily His plans.

Here is an old Yiddish proverb: "Men plan and God laughs."

[I first saw the above Yiddish phrase on *Digihitch.com*. A guy named Redford had it on his posts.]

Never Bring a Knife to a Gun Fight

A few years ago I was walking north out of Jackson, Wyoming. This guy pulled over and picked me up. He drove me all the way to Dubois. We had a good talk.

"You're the first hitchhiker I have picked up in fifteen years," he said.

"Why's that?" I asked.

"Fifteen years ago I was driving through Rawlins and I saw this hitchhiker. I pulled over and he got in my pickup. We're driving down the road and he pulls out this knife and starts cleaning his fingernails. Then he asks me how much money I have in my wallet. I think I said around fifty bucks. I didn't think much of what he said at the time. Then the hitchhiker said, 'Why don't you just hand over your wallet and nothing will happen to you.' Then I realized that this was a robbery."

"Did he threaten you with the knife?" I asked.

"No. He just kept cleaning his fingernails. So I handed over my wallet to him and he told me to drop him off at the next exit. We drove a short while further and I dropped him off at this exit. The hitchhiker got out of my pickup and I drove off and stopped a hundred yards from where the hitchhiker was standing. I could see in my rearview mirror that the hitchhiker was looking at me when I stopped the pickup."

I was also curious as to why he stopped his pickup there on the shoulder.

"I reached behind my seat and took out my handgun. I put the pickup in reverse and drove back to the hitchhiker. The hitchhiker had a surprised look on his face. I rolled down the passenger window, pointed my gun at him and said, 'Give me back my wallet.' The hitchhiker gave me back my wallet. Then I said, 'Now give me YOUR wallet.'"

I started laughing.

"The hitchhiker gave me his wallet and I drove off. And that's why I haven't picked up a hitchhiker in fifteen years."

The moral of the story: Never bring a knife to a gunfight and never clean your fingernails in someone else's pickup.

Someone Gets a Free Gas Can

Today I hitchhiked from Riverton to Farson to Pinedale, Wyoming. The guy who gave me the ride picked me up a couple of miles south of Riverton. He was coming from the oil fields in North Dakota. We had a good talk.

He told me that he knew this older guy who was driving his pickup on I-80 near Wamsutter, Wyoming some years ago. This guy saw someone walking down the interstate with a red gas can, so he pulled over to give him a ride. The hitchhiker got in the cab of the pickup and pulled a gun on him. He told the driver to give him his money, his cell phone and his pickup. The driver was forced out of his pickup at gunpoint. The hitchhiker gave the driver the gas can and drove off.

At least the driver got a free gas can.

As we drove north to Pinedale, the guy giving me a ride asked me if I carried any protection. I said, no, I carried no gun or knife. He said that a flare would make a good defensive weapon. If the sulphur didn't smoke the guy out, you could burn him with it.

I don't carry any weapons; I put my trust in God (read Psalm 91).

Goodbye, Las Vegas

The last time I hitchhiked through Las Vegas was in December of 2005. I haven't been back since. Goodbye, Las Vegas (Unreal City).

Goodbye, Las Vegas
By Tim Shey

"Unreal City,
Under the brown fog of a winter dawn,
A crowd flowed over London Bridge, so many,
I had not thought death had undone so many."
"He who was living is now dead
We who were living are now dying"
"Falling towers
Jerusalem Athens Alexandria
Vienna London
Unreal"

--T.S. Eliot

"Goodbye, Las Vegas"

Desert jackals
Run to their destruction
Hollow eyes see nothing
Behind shades of glass
Painted Jezebel faces
Unrecognized by man
Mourning becomes electric
As piercing city lights
Rape the virgin night
This place never sleeps
And never awakes from death
Black Jack table bait
Roll-the-dice breath
Throw your money down
This is casino heaven
Idolatry never felt so good
This harlot language doesn't speak
Straw fires always burn fast
I see the Prophet Jeremiah weeping
Over a people brought down to bankruptcy
By a Queen, a King and three Aces
A hitchhiker wanders hardened streets
With his burden on his back
This is the heart of darkness
Lifeless buildings built with foolish gold
I see Sodom burning
And bodies turned to ash
They were very fluent
In arrogance, pride, adultery
And enviropaganspeak
You have sold your soul to Satan
Do you remember Noah's Flood?

The City of David was sacked by Romans
And America by Marxist-Darwin thugs
The Stranger leaves the graveyard
And the stench of Vegas Past
And hitches a ride to Barstow
Across the relentless Mohave
On Interstate Fifteen

The Strangest Thing I Ever Saw

High Plains Drifter: A Hitchhiking Journey Across America

By Tim Shey

Excerpt from Chapter Five:

Psalm 18: 19: "He brought me forth also into a large place; he delivered me, because he delighted in me."

In May of 1997, I hitchhiked west towards Nebraska. I have always loved going through Nebraska. In all my travels, I believe that the people of Nebraska and Texas were by far the best people I have ever met. Nebraska was in my comfort zone. Whenever I had been wandering out west and came out of Wyoming and into Nebraska, I felt that I was back on my home turf. The people of Nebraska are gold, silver and precious stones.

I got some good rides all the way to Osmond, Nebraska. It was getting close to sundown when this guy driving a tractor picked me up.

"You can sit on the fender if you want," he said.

"Sounds good to me," I said.

He drove me to Plainview and we talked about the things of God and the Bible. He asked me what I was doing. I told him that I just quit my job and thought I would hitchhike by faith and see where God would take me. He offered me a job right there. He had his own construction company and lived on a farm with his wife and kids. I told him I would love to work for him, but that God was calling me out west for some reason. We stopped in Plainview and we shook hands. I hopped off the tractor and I got a motel room.

The next morning this tractor-trailer picked me up.

"I'm going all the way to western South Dakota," he said. "I got five drops: three in Nebraska and two in South Dakota."

We stopped at three places in Nebraska and I helped unload his van--he was hauling some small trees and shrubs. We got to a truck stop near Kadoka, South Dakota and he told me he would buy me some supper.

We were eating supper when he looked at me and said, "You know, right before I picked you up I saw this man pointing at you. It was like he was telling me to pick you up."

"What?" I exclaimed. "I didn't see anybody out there. I was alone."

"I saw him plain as day. When I picked you up I didn't see him anymore."

I was flabbergasted. So I sat there and wondered and looked out the window and asked him, "Do you think he was an angel?"

"He must've. It was the strangest thing I ever saw."

On our trip we talked a lot about the Word of God and certain preachers on TV. He lived in Sioux City and was very well self-educated. I enjoyed talking with him. After supper he said he was going back to the sleeper and get some sleep. I took a long walk--for two or three miles--out in the country. Lots of grassland; it was beautiful.

I walked back to the truck and the trucker was sound asleep. He had a double-decker sleeper, so I got in the top bunk and turned on the VCR. From midnight till four in the morning I watched two films. The first film was *The Professional*--it was about the life of a mafia hit man and a twelve-year-old girl named Matilda. It was very good. I forget the other film.

The next day we stopped in Rapid City and Spearfish and we unloaded his truck. He bought me breakfast and I hit the road.

At a Cafe in Merriman, Nebraska

Yesterday I hitchhiked from Valentine to Merriman. I phoned Steve and he drove to town and took me and his son, Will, to a local cafe for dinner. Steve and his wife Carol have a cattle ranch thirteen miles from Merriman. Their son, Brock, and their daughter, Tiffany, also work on the ranch. Steve had picked me up hitchhiking back in 2006, so I thought I would stop by and say hello.

Steve, Will and I sat down at a table and ordered something to eat. A few minutes later, this other guy walked in and sat down with us. He looked like he was in his late 50s. His name was Chuck.

We talked about various things: ranching, hitchhiking, politics. Chuck then started talking about his experience in the Vietnam War. He was a Navy SEAL that had graduated from BUD/S (Basic Underwater Demolition/SEAL) Training in 1972. Chuck talked at length about some of his firefights in the jungles of Southeast Asia. He said that the average life expectancy of a lieutenant in Vietnam was eleven minutes. Chuck was once shot out of a tree by an RPG (rocket-propelled grenade); he was providing covering fire for his team when the explosion of the grenade knocked him out of the tree. He had intense, penetrating eyes; it looked like he had been to hell and back.

I asked Chuck if he had seen the film *We Were Soldiers* and if it was a realistic account of combat in Vietnam. He said that he had seen the film and that it was very realistic. Chuck said that he had met Hal Moore (the author of the book *We Were Soldiers*) and thought that he was the best officer in Vietnam. I believe Moore had retired as a general in the U.S. Army.

Chuck had a son who fought recently in Afghanistan. He was an Airborne Ranger. Chuck talked a little about his son's combat experiences on the border of Pakistan and Afghanistan.

Some people think that the Navy SEALs are the best elite warriors in the world and some people think that the British SAS are the best. I asked Chuck if he had ever met any British SAS; he said that he had met a few. I could tell that Chuck knew where I was going with this: are the SAS the best warriors in the world? Chuck told me that the Israeli Special Forces were "deadly"; he had absolute respect for them and for Mossad (Israeli Intelligence). He said that the Israeli Special Forces were the best elite soldiers on the planet.

We finished our dinner and I shook Chuck's hand. It was a great honor to talk with a U.S. Navy SEAL.

I remember watching a documentary on President Harry Truman. Since a child, Truman had to wear glasses--he was pretty much blind without them. In a World War I photo of Captain Harry Truman, he had his glasses off. The commentator of the documentary said that Harry Truman had eyes of steel. Chuck, the Vietnam Veteran, had eyes of steel.

I stayed overnight at Steve and Carol's ranch. Steve, Carol, Tiffany and myself had excellent fellowship at the supper table. Tiffany was hoping to get into a Christian college in North Carolina. I told them a number of my stories of hitchhiking around the United States. They have a beautiful ranch in the Sand Hills of Nebraska. I was grateful to have met Steve's family. I also met Steve's dad and step-mom. Steve's dad writes for three newspapers in Nebraska and one in South Dakota. Steve's dad gave me a copy of a booklet that he had published; these were newspaper articles that were published during the previous year.

Right now I am in Chadron. I may be heading south to Alliance tomorrow.

The Things I Carry

I weighed my backpack about a week ago and it weighed 56 pounds. It is a *North Face* backpack. My friends bought it for me at a garage sale in Jackson, Wyoming in October of 2006; they paid 50 bucks. It has a lot of duct tape and gorilla tape on it. In 2009 I voted gorilla tape my Most Valuable Player.

This is what I carry in my backpack:

1 summer sleeping bag
1 *Coleman* winter sleeping bag (rated at 10 degrees F.)
1 two-man tent
1 *Muleskins* winter coat
1 *Cabelas* hooded sweat shirt
1 pair *Billabong* shorts
1 insulated flannel shirt
An extra baseball cap
1 compact pillow
1 roll toilet paper
1 package *Bic* shavers
2 stocking caps (1 full mask)

1 pair winter gloves

1 *Duracell* flashlight/radio (This is one of the best things ever given to me on the road. You don't need batteries; there is a handle you use to wind it up and recharge it. This Canadian Army veteran of Afghanistan picked me up outside of Lolo, Montana and gave me a ride to Lolo Pass. He said the flashlight was brand new. He was from Alberta, Canada.)

2 water bottles (1 liter each)
1 can opener
1 pair reading glasses
1 watch
Shaving kit
2 *Bic* lighters
Ear warmers

Leatherman all-purpose tool
Various articles of clothing (socks, underwear, etc.)
1 compact King James Bible
A *Mead* folder that holds:
A road atlas
A pocket-sized daily planner/calendar for 2010
3 pens
A 100-page notebook
A folder that holds some photocopies (11 pages) of *Milton and the English Revolution* by Christopher Hill
A copy of my seven-year book contract with *PublishAmerica*
1 spoon

A Week in the Life of a Hitchhiker

In the past week, I hitchhiked from Helena, Montana to Dayton, Washington. The ride from Helena took me to Missoula. This guy's name was Harry and he came from the Fort Peck Indian Reservation in northeast Montana. Harry was from the Assiniboine Tribe; we had a good talk. I told him that I was a Christian. He knew very little about Christianity. I told him about my faith in Jesus and that he should read the Gospel of John in the New Testament. I think he said that someone gave him a copy of the New Testament some years ago.

It had been snowing that morning when I left Helena and there was some slush on I-90. Harry was going 85 miles per hour when he hit a patch of slush. All of a sudden, we were going sideways down the interstate. Then we went sideways down into the median (I thought we were going to roll his van over) and continued going sideways into the next lane into oncoming traffic. This big tractor-trailer was bearing down on us and I thought we were going to get T-boned by the tractor-trailer when, all of a sudden, the van straightened itself out. Harry took control and we drove on the shoulder to the next exit. That happened near Clinton, Montana.

It was quite a rush for at least several seconds. It all happened so quickly. Harry and I looked at each other and heaved a sigh of relief. Harry said that my God saved us. I said, Praise the Lord!

Harry was in a hurry to get to this hospital in Missoula; he had injured his back getting bucked off of a horse during his rodeo days. We went to this hospital where they gave him some shots in his back. I sat and talked with Harry as he lay in the bed. The nurses thought it was pretty funny that he had picked up a hitchhiker.

After the hospital, Harry took me to his relations' place in Missoula and I slept on the floor that night. The next morning, his nephew drove me to Lolo where I started walking west on U.S. 12.

I walked a couple of miles or so and this married couple in a vehicle pulled over. They were Michael and Sandy and we had some excellent fellowship--they were really in tune with the Holy Ghost.

We drove to a cabin that they had rented and had a powerful prayer meeting. The demons were manifesting in Michael as I commanded them to come out. We later had breakfast at a local bar/restaurant and then headed back to Clinton where I stayed at their place for the night. The next day Michael drove me over Lolo Pass to Lochsa Lodge and dropped me off. Then I walked a few miles and got a ride to Kooskia, Idaho.

From Kooskia I got a ride to Kim and Pat Hosking's place between Stites and Harpster. I met Kim and Pat while I was hitchhiking on U.S. 12 near Lolo, Montana in 2004. Kim builds wood furniture and has a portable band saw, so he can cut up logs into boards.

I hadn't seen Kim and Pat in a year. They let me stay for five nights. I helped Kim cut some white pine, yellow pine and red fir logs on his band saw. Pat was doing some editing on her book *The Lion's Roar* (her pen name is Margaret Hosking).

Yesterday, I hitchhiked from Kim and Pat's place to Kamiah. From Kamiah I hitchhiked to Lewiston, walked across Clarkston, Washington and got a ride to Dayton where I will be staying with Gene and Tanya.

I met Gene and Tanya back in October of 2008. Gene and his son picked me up hitchhiking in Walla Walla and took me home. They asked me to speak at their church in Dayton. Tanya gave me a sleeping bag back in 2008 which I still have.

Gene and Tanya and I have had some real good fellowship. I may be here for a few nights and then head into Oregon.

It rained last night; the skies are overcast now. I got a real good sunburn on my neck and arms after working with Kim on the band saw. It is a real blessing to be out of the sun for a few days.

The Pacific Ocean

This morning I hitchhiked from Philomath to Newport, Oregon on U.S. 20. It is a beautiful, partly-cloudy day here on the beach just west of Newport. There is a constant roar from the waves of the Pacific crashing onto the sandy beach. I found a nice rock with a flat top to do my writing. There is a grassy cliff behind me and a vast expanse of blue before me.

The last time I was in Newport was back in 2001. I had hitchhiked from Iowa to the coast of Oregon travelling mostly on U.S. 20.

I remember I was hitchhiking in Virginia in the late 1990s and this guy picked me up. He knew a guy who had been all over the world--this guy said that the most beautiful place on the planet was southwest Virginia; he also said that the most beautiful coastline in the world was the coast of Oregon.

I used to live two blocks from the Pacific Ocean while I was living in Venice, California. I lived in Venice on Howland Canal for three months in the spring of 1984. At that time, I was working with *The Horse and Bird Press* of Los Angeles: this press published the poetry of Carolyn Kleefeld. I sold books in New Hampshire, Vermont and California (I was not a very good salesman). I house-sat for Patricia Karahan who was the publisher of *The Horse and Bird Press*. Patricia had gone to Greece and Spain on vacation--that is how I ended up in Venice.

I was told by someone who lived in Los Angeles that LA was the most stressful place to live in the United States. I would have to agree. Every Wednesday I would drive to Century City and pick up the mail for *The Horse and Bird Press*. When I got back, I would have a splitting headache: the traffic, the people, the air pollution all contributed to the stress.

The first week that I was in Venice, I had a persistent sore throat. It was probably from the smog. So I would drink a quart of orange juice every day. My sore throat disappeared.

Usually, every night I would walk the two blocks to the beach just to sit on the sand and listen to the pleasant roar of the waves hitting the beach--I guess it was my therapy for living in such a stressful city.

I remember seeing this guy walking around the boardwalk in Venice in a white robe--he was also barefoot (but I don't think he was with the Discalced Carmelites). I thought that he was a Hare Krishna follower. So I walked up to him and asked him about his beliefs. He told me that he was a Christian and that he walked in faith. I told him that I had been a Christian for two years. I asked him if he needed a place to stay for the night. So he stayed at my place on Howland Canal for the night.

I made him some soup and sandwiches and then we had a good talk about the things of God that evening. He told me that he had spent some time in Italy: the people there thought that he was Francis of Assisi. I had a copy of Thomas Merton's *The Wisdom of the Desert*, so I gave it to him. He was very grateful.

He slept on the living room floor that night and left the next morning. That was probably in April of 1984.

Today, there is a gentle breeze coming in from the ocean. I am glad that it is not raining. It is sunny: there are clouds, but they are high-altitude clouds. There are people walking on the beach.

There is a lighthouse to the north--it is around three miles as the crow flies from where I am sitting. There are a few sea gulls gliding around just above the cliff. There are three people flying kites to the north of me. I see a ship in the distance on the horizon to the port side of me (I have always wanted to say that). So that means that the lighthouse is to the starboard (now I am starting to think of *Moby Dick* by Herman Melville). This reminds me of the time I hitchhiked up Highway 1 on the coast of California back in the late 1990s: I slept in this grassy field near this lighthouse--I believe it was Point Sur. California also has a very beautiful coastline.

Speaking of the coast of California. I hitchhiked from Nebraska to California back in April of 1983. I stayed with a friend in Big Sur for a week. I then hitchhiked down Highway 1 to a place near Santa Lucia (I don't think this town exists anymore). There was this Camaldolese

Monastery near Santa Lucia; the monks let me stay there for three nights--I had my own hermit cell. During that trip, this man and woman picked me up and told me that a friend of theirs had a dream about an earthquake that was going to hit California, so she flew to Thailand. Within a week or so of me hearing about this, an earthquake hit the Coalinga area of California (2 May 1983; 6.5 on the Richter Scale).

It has been a very blessed trip from Montana to Oregon. I am breathing and hearing and seeing God's Creation here where the Pacific meets the edge of the Universe. "Breathe, arch and Original Breath" (Gerard Manley Hopkins). The Presence of God has been very strong in the past few days. I thank the Lord for bringing me back to the Pacific Ocean.

God willing, I will head south from Newport on U.S. 101.

Picked You Up On the Road

This morning I hitchhiked from Cameron, Montana to Ashton, Idaho. I even got a ride from an Idaho deputy sheriff (from near Henry's Lake to Ashton).

The deputy sheriff stopped and asked, "You're walking in this shit?" The wind was blowing hard and there was a little snow falling.

I said, "I'm hitchhiking."

"You know that it is illegal to hitchhike in Idaho, don't you?"

"Yeah, I know."

So he checked my ID and had me put my backpack in the trunk of his car. He then drove me to Ashton.

When I got to Ashton, I went to the library to check my email. I got a real nice email from a lady that picked me up south of Three Forks, Montana a few days ago. She and her daughter gave me a ride to Ennis and dropped me off at the Exxon gas station. Here is her email:

"Hey Tim,

I wanted to say hi. And I would really like to hear more of what you have to say. You are the most alive person I have met in a long time. Maybe ever.

We are going to Norris hot springs Tues. eve. if you would like to join us. Or I would fix dinner--if you bring the conversation. We live in Pony. If you are still close.

Thanks, take care, peace, Karry"

After Ashton, I will probably mosey on down to Driggs and Victor, Idaho and look up some friends. It is still snowing here in Ashton.

I had been staying with some friends in Ennis for a couple of nights this weekend. My friend dropped me off at the Exxon Station in Ennis and then I started walking south. I didn't get any rides that afternoon, so I walked to Cameron and camped out at the post office last night.

*****

Back in Montana

A few days ago, I hitchhiked from Victor, Idaho to Ennis, Montana. When I left Victor, it was around 10 degrees F; when I got to Ennis, it was close to 40 degrees--it really warmed up. I was at the library in Ennis for an hour or two and then I hit the road.

I got a couple of rides to Harrison. The guy who gave me a ride to Harrison let me use his cell phone, so I was able to phone Karry. She said she was on her way. I got dropped off near the post office in Harrison and waited ten minutes and Karry picked me up.

Karry had picked me up a couple of weeks earlier when I was walking south of Three Forks. She had emailed me to say that I was welcome to stop by her place any time. I stayed with her and her daughter for two nights in a small town in the mountains--Pony, Montana.

Karry is close to my age and she is an artist. She paints in watercolors, oils and has done sculptures and murals. I liked her work a lot. A long time ago, I was really into Vincent Van Gogh and other impressionist and post-impressionist painters.

A week or so ago, this one guy picked me up and told me that there is a difference between being a painter and an artist. So I asked Karry if she was a painter or an artist; she said that she was an artist. She has some of her paintings in an art gallery in Ennis.

Her son, Adam, wrote a children's book recently and Karry did the illustrations. The title is *Wandering Bard* by Adam Root, illustrated by Karry Hesla.

Just got dropped here in Belgrade, Montana. I may have a place to stay here for the night. God willing, I will be heading for western Kansas.

Good Karma

High Plains Drifter: A Hitchhiking Journey Across America
By Tim Shey
Excerpt from Chapter Nine:

I walked down the road for a mile or two, and this truck driver saw me and took me to a truck stop in Tennessee. He preached to me in the power of the Holy Ghost all the way to Knoxville. At Knoxville this guy picked me up and took me all the way to Fort Smith, Arkansas. We stopped at a truck stop in West Memphis, Arkansas around midnight.

This twelve or thirteen-year-old kid walked up to me and asked me to give him a ride to California. I told him he had to ask the driver. The driver got pretty upset and told the kid to go home to his parents. The kid walked off. He was pretty young--it would be dangerous for him to be on the road. I sure wasn't thinking about hitchhiking when I was that age. Maybe he didn't have much of a home life.

I got rides from Fort Smith to Amarillo to Lubbock and then to Roswell. From Roswell I got a great ride from a truck driver all the way to Antonito, Colorado. I got to Alamosa and this lady named Nancy picked me up and she gave me a sandwich and something to drink. She let me off north of Alamosa and then I got rides to Salida and then to Canon City. There I slept under a doublewide home in a sales lot.

The next day I hit Pueblo, then Walsenburg and headed back west on US 160. I got into Del Norte and I went to the sheriff's office to see if someone would put me up for the night. A local church put me up in a motel. Nancy told me she lived in Del Norte, so I went to look her up. She lived close to the sheriff's office and she was surprised to see me. We talked for a while and then she drove me to Pagosa Springs.

I got rides to Durango and Cortez and then I was dropped off near Dove Creek where I slept in somebody's machine shed. It rained hard that night and I was grateful to be warm and dry. I woke up around four in the morning and began walking down the main street of Dove Creek. I found an old Kenworth or Peterbilt tractor and crawled into the sleeper and slept for two or three hours. The mattress of the sleeper was more comfortable than the dirt floor of the machine shed.

It was now daylight and I thought I had better get out of the truck before somebody walked up to it and drove off with it. That reminded me of the time back in July 1980 when I hopped a freight train in Fremont, Nebraska and I rode it all the way to a place called Chapman--near Grand Island. This cop saw me riding on a flatcar and unfortunately the train stopped. The cop drove his car to where I was sitting and told me to get off the flatcar. So I jumped off the train and got in the police car. To make a long story short, the cop dropped me off at the county line and I had to walk six miles that night to the next town. The name of the town was Duncan and, by the time I got to Duncan, I had developed a pretty bad attitude. I was tired, thirsty and I got caught riding a freight train--I was not a happy camper. Anyway, I saw this pickup parked by the railroad tracks and slept in the cab that night. I woke up and walked to US 30 and stood there thumbing for a ride to Columbus. About a half hour later, I saw this guy walk up to the pickup and drive off in it. Sometimes it pays to get up early in the morning.

From Dove Creek I walked to a truck stop, got something to eat and walked several miles west. A truck driver picked me up and we drove through Utah up to Salt Lake City and then east to Wyoming. We drove north of Rock Springs and unloaded his trailer at a gas field. We then drove to northern Utah and loaded his trailer with steel. We drove back to near Farmington, New Mexico to his ranch where he and his dad lived. I stayed there a few days and helped do cleanup around the place. We then drove to Albuquerque where he dropped me off.

From Albuquerque I headed west on I-40 and got a motel room in Grants. From there I headed south and west on Highway 53 and then south on US 191 near the Zuni Indian Reservation in Arizona. I walked several miles and found this abandoned building by the side of the road. I jumped the fence and walked behind the building about fifty yards and camped there that night. I slept there in my sleeping bag and listened to the coyotes yelp and howl.

The next day I got to St. Johns, Show Low and then to Globe where I slept out in some bushes on a hillside. The next morning I got a few rides to downtown Phoenix and then I started walking. I must have walked ten or fifteen miles and slept somewhere off the road someplace. The next morning I reached Litchfield Road and got a couple of rides to Blythe, California.

It was a hundred and ten degrees in Blythe. In Phoenix the day before, it was a hundred degrees-I stopped several times to fill up my water bottle. After an hour wait, this guy in a van picked me up. I got in the van and looked at the guy--he was rubbing the back of his head.

"What's wrong with your head?" I asked.

"I got robbed by a hitchhiker," he said.

"What?" I exclaimed.

"Yup. He hit me over the head at a rest area down the road and stole four hundred dollars I had on me."

"Then why did you ever pick me up?" I asked completely dumbfounded.

"I needed all the good karma I could get."

I sat there in disbelief as we drove up US 95 to Vidal. He was hoping that I had some money on me to help pay for gas. I told him I was sorry, but that I was broke. We talked for a while and then he casually mentioned that he had a box of *Poptarts* in the back seat. I hadn't eaten in fifty-two hours; those were the best *Poptarts* I ever had.

We drove to Vidal and we stopped at a gas station. There I talked with the kid that worked behind the counter. I told him that a hitchhiker robbed this guy, and that he was trying to sell some camping equipment so that he could buy some gas and get back to Ridgecrest. He said, no problem. He bought thirty dollars' worth of equipment and we were off.

We drove north to Needles and headed west on I-40. Somewhere near Ludlow we stopped at a truck stop. He slept in the van and I slept on the ground. The next day we made it to Ridgecrest and I headed north on US 395.

Miguel the Chef

Yesterday I was walking south of Four Corners, Montana (west of Bozeman) and this guy picked me up. His name was Miguel and he was driving to Big Sky. He was born in Santa Monica, California; his parents were from Spain and England, respectively. He did live for a time in Newcastle in northern England; he had a slight accent.

We had a good conversation. Miguel is a chef and he cooks for people in their homes. When I met him, he was going to Justin Timberlake's house to cook for twenty people. I guess Miguel has cooked for some very wealthy people at the Yellowstone Club in Big Sky. He has cooked for Bill Gates and Warren Buffet.

One time Bill Gates asked Miguel to cook for he and his friends at his house at the Yellowstone Club. Miguel showed up and Bill Gates handed him a hundred-dollar bill and told him that he didn't have to cook; they were going to McDonalds to get some hamburgers. So Bill Gates and his friends took Gates' helicopter and flew through the canyon that goes from Big Sky to just south of Four Corners, flew to Belgrade to the airport, hopped in a car and drove to McDonalds. They got their food, hopped back in their car, drove back to the airport, hopped back in the helicopter and flew back to Big Sky.

I guess that is all fine and dandy, but if Bill Gates EVER borrows MY helicopter without MY permission so that HE can go to McDonalds, I might get a bit grumpy.

Now that I have heard of everything, I can die happy.

Without McDonalds we will be a people no more.

*****

A few years ago, I was hitchhiking north through Big Sky to Bozeman and this guy picked me up. He told me a story about Bill Gates. One time Bill Gates' car broke down and he was trying to change his tire. This local guy stopped by and asked Gates if he needed any help. Gates told him that that would be great. So the local guy helped change the tire and then started to walk back to his car to drive off.

Bill Gates said, "Can't I give you some money for helping me out?"

The local guy said, "That's all right. Just help out the next guy who's broke down."

Gates said, "Do you know who I am?"

The local guy said, "You look familiar."

Gates said, "I'm Bill Gates." ("Bill Gates" is coded language for "The Wealthiest Guy on the Planet" or "I AM Microsoft")

So Gates told the local guy to follow him to Bozeman.

They drove to Bozeman to a car dealership. Bill Gates bought the guy a brand new SUV.

*****

P.S. After Miguel dropped me off at the gas station in Big Sky, he gave me some money; I was very grateful because I was broke. I bought a sandwich and then hit the road.

A Ride on the Reservation

This morning I was walking a mile or two south of Mission, South Dakota on U.S. 83 when this vehicle pulled over to pick me up. This guy was from the Lakota Tribe on the Rosebud Reservation. It was really windy and cold, so I was grateful to be in a heated vehicle for a short while. I believe it was below zero with the wind chill.

This guy told me that he had a dream a short while ago and in the dream he saw a guy walking down the road, so he picked him up. When he saw me walking down the road, he had to pick me up.

I told him that I have had a lot of dreams from the Lord and that some of these dreams have come true. He then told me that back in 2000 he had a dream about an airplane that crashed into two tall buildings. I said that the Lord does show things to people in dreams or He warns people in dreams pertaining to future events.

The Lord can give dreams to believers and unbelievers. The Lord gave dreams to Pharaoh and Nebuchadnezzar. He gave dreams to Jews and Christians. I have heard that the Lord is giving dreams to Muslims about Jesus and they are getting saved. Praise the Lord!

God is sovereign; He rules in the affairs of men.

A few days ago, I hitchhiked from Bozeman, Montana to Bismarck, North Dakota. Two days ago, I hitchhiked from Bismarck to Pierre, South Dakota. Today I made it to Valentine, Nebraska. Later this afternoon, I may visit a friend who lives just east of Valentine.

Just remembered: yesterday I was walking a few miles east of Bismarck, North Dakota on I-94 when this guy pulled over to give me a ride. He dropped me off at Sterling, North Dakota. He told me that he picks up every hitchhiker he sees because it might be Jesus walking down the road.

It's a Small World

Yesterday I was walking east on U.S. 20 between Bassett and Stuart, Nebraska when this car pulled over to give me a ride. This guy's name was Shawn and he was going to Atkinson on an errand. We got to talking and he just got back from a mission trip to Mexico. Shawn used to be a pastor at a few churches. He recently lived in the Star Valley area of western Wyoming. He now lived in Valentine, Nebraska with his wife Theresa.

After Atkinson, we drove to Ainsworth to pick up his wife. We stayed at their friends' place for supper and then drove west of Ainsworth to this farm to see a couple that they knew. We walked to the house and the man motioned for us to come inside. I looked at the man and he looked familiar. His name was Greg and his wife was Marla.

We talked for a while and Shawn told Greg and Marla that he had picked me up on the road earlier that day. I think Shawn then asked Greg if he had ever picked up any hitchhikers. Greg said that he and his wife picked up this hitchhiker in Idaho four or five years ago and that the hitchhiker had written a book. They dropped the hitchhiker off in Missoula, Montana.

Greg then said that the hitchhiker sent him a copy of his book. He searched for a short while and then produced the book [typescript]. It was my book! (*High Plains Drifter*)

It was a photocopy that this lady in Lewiston, Idaho had sent to them. She picked me up hitchhiking in the fall of 2004 and told me to give me a floppy disk of my book and that she would make some photocopies and then send it to anyone I wanted. She owned a print shop in Lewiston.

I told Greg that he probably picked me up on U.S. 12 somewhere between Kooskia and Lolo Pass, Idaho in the fall of 2004. We talked about it some more and I believe he picked me up at a gas station at Lowell or Syringa, Idaho.

We stayed at Greg and Marla's place for an hour or so and had some excellent fellowship.

It's a small world.

Branding Calves and the California Outback

A week ago I hitchhiked from Mount Vernon, Oregon to Cedarville, California. John and Susie were happy to see me again. I had been gone for a week.

A couple of days ago, John, Susie, four of their friends and I drove from Cedarville to a ranch between Burney and Fall River Mills, California. The ranch is forty miles north of Mount Lassen (Lassen Volcanic National Park). John and Susie have a cow-calf herd there; the calves needed branding.

Two guys on horses would rope the calf--one guy roped the calf's neck and the other guy would rope the hind legs. They would drag the roped calf close to where we were standing and then John and I would grab the calf and flip it on its side. One of us would loop the rope around the calf's front legs just under the hooves and make sure it was drawn up tight. We would do the same thing to the back legs. The guys on the horses would then take up the slack in the ropes, so that the calf would not move.

John would then notch the ears and castrate the bull calves. Susie and her daughter would inject the calves with Vitamin E, there was an 8-Way shot, and there were shots for pneumonia, scours and pinkeye--they had two syringes each. I would then take the hot branding iron and brand the calves high on the flank. We used the same brand for all of the calves except three.

We ended up branding around eighty head. We then walked to the house, had a steak dinner and then headed back to Cedarville. We left Cedarville at eight in the morning; we got back at six.

This morning John and I drove out to a piece of ground they own. It is around eighteen miles north and east of Cedarville. There were lots of sagebrush and Juniper trees. This is high desert country (Cedarville is 4600 feet in elevation): they call this the California Outback. This country is much different from the Fall River Mills neighborhood which has a lot more trees, good pasture and some farm ground.

We dug (or tried to dig) a couple of post holes on this ridge--we brought a couple of rail road ties with us. John wants to build a gate that leads out to his property. We also brought a couple of gate panels. There were a lot of rocks in the ground, so we weren't able to dig down very far. We will come back later and pour concrete around the end posts. After the concrete is set up, we will hang the gate panels on the end posts.

The Nevada state line was a mile away on another ridge. It was pretty muddy driving out there. We had to put John's rig into four-wheel drive. They have had a lot of rain and snow this spring. It is supposed to rain off and on for the next few days.

Greensburg, Kansas

I am standing on the steps of the courthouse in Greensburg, Kansas. I just hitchhiked from Pratt to Greensburg this afternoon on U.S. 54. A year ago (4 May 2007) a tornado hit Greensburg; it looks like it totally destroyed eighty per cent of the town. Houses were taken off of their foundations, lots of trees were uprooted and there are still pieces of metal embedded in many of the trees that are left standing. Locals told me that the tornado was two miles wide. I have never seen devastation like I have seen here in Greensburg. The people of Greensburg have done an excellent job in cleaning up their town; you see brand new homes going up everywhere.

I walked past a *CBS News* trailer. President Bush is going to be in town this weekend to speak at the Greensburg High School Graduation Commencement. I am sure there will be a lot of media in town for the President's visit.

This past week I stayed at Lawrence and Cheryl's place in rural Stafford, Kansas. I met Cheryl and her daughter, Jessica, and Jessica's husband, Grisha, six years ago when I was hitchhiking through St. John. My home base from November 2001 to August 2002 was St. John. I knew several people in St. John and would stay at their homes whenever I was passing through. Those people no longer live in St. John. So it definitely was the Lord's will to go to Stafford.

While I was staying at Lawrence and Cheryl's place, a lady named Connie phoned me and asked me to speak at the First Baptist Church in Stafford on Sunday. So I preached on Acts Chapter 10 and on obedience to the Lord. The Lord really blessed me for preaching at First Baptist. The congregation also gave me a generous offering, so I was able to get a motel in Pratt last night, I got a haircut this morning and I made photocopies of *High Plains Drifter* and *Dreams from the Lord* and mailed them to Lawrence and Cheryl. Cheryl is not into computers and the Internet, so now she can read the photocopies instead of using the Internet.

Since I saw Jessica and Grisha last (2002), they have had four kids. The oldest is Jesse (5 years) and the second oldest is David (3 years);

they also have baby twins-a boy and a girl. Jessica later told me that when she was praying with the kids before bedtime, David said this: "Dear God, thank you that Mr. Tim is not dead. If Mr. Tim wants a toy, please give him a toy. If Mr. Tim needs a car, please give him a car." I thought it was so funny.

It was a very blessed week for me. I spent part of three days pruning the trees and cleaning up broken limbs around Lawrence and Cheryl's place--there was an ice storm in January. They let me use the car, so I was able to go to the library in Stafford and in St. John to get some work done on my website. Lawrence and Cheryl have a beautiful, peaceful place out in the country; I enjoyed taking walks down the gravel road and in the fields with their four pet dogs. It looks like the winter wheat is doing very well--they must've had plenty of snow this winter.

The courthouse here in Greensburg is still standing, but it doesn't look like it is being used at this time. To the west and south of here (the corner of Florida and Oak Street) is where most of the devastation happened. Someone told me that eleven people died because of the tornado. On the corner of Florida and Main Street, there is a lone, brick building standing. All around this building nothing was left, just rubble. It looks like this building was at "ground zero."

Looks like I will head west to Garden City. From Garden I will then mosey on up north into Nebraska on U.S. 83. It has been a beautiful, breezy day. It is nice to be in this part of Kansas again. When I hitchhiked back to St. John and Stafford, it felt like I was coming home.

Rock Springs, Wyoming to Barstow, California

A little over a week ago, I was walking south of Rock Springs, Wyoming on U.S. 191 when this vehicle pulled over. This guy was Pastor Rich Carlson and we had some pretty intense fellowship for the fourteen miles of a ride he gave me. Pastor Carlson prayed for me and gave me a little money for the road.

I walked for three or four miles and this truck driver picked me up. I got in the cab of the tractor-trailer and he looked at me and said something like, "Aren't you done hitchhiking all over the place and spreading the Word?" I recognized him right away; he had picked me up over a year ago on the same stretch of road. He later told me that he was a pagan. We didn't talk about much. Later, just before he dropped me off, we talked a little about Dostoyevsky, and things Russian; he told me how to properly pronounce "Dostoyevsky" and "Karamazov"--at least, how the Russians pronounce it.

He dropped me off in Vernal, Utah where I slept in the post office that night. It got down to maybe ten degrees that night, so it was nice to sleep in a warm place for the night. Somebody phoned the police that I was sleeping in the post office because I was woke up around eleven that night by a couple of police officers. They checked my ID and let me sleep in the post office that night. I told them that I would leave as soon as possible the next morning.

I then hitchhiked from Vernal and made it to Helper, Utah where I stayed at a shelter for three nights. From Helper, I made it to Mexican Hat, Utah where I slept in a junked pickup camper for the night. The next day, I got good rides through Kayenta and Tuba City to Flagstaff, Arizona where a guy named Tim picked me up.

He was driving one of those big motor homes from Iowa to southern California. Tim said that the motor home he was driving was worth $270,000.00. He told me that he was in a car crash up in Iowa that involved the car he was riding in and two tractor-trailers. He said that this tractor-trailer ahead of him on the interstate had jack-knifed and the guy driving their car lost control and the car began spinning around and this tractor-trailer came up from behind them and it looked like they were going to be crushed. If I remember right, Tim said that this powerful force threw their car into the ditch. Tim said that his mind still had not accepted the fact that he was still alive. He thought for sure that they were going to be crushed to death by the two tractor-trailers. I told him that the Lord preserved him in that crash because He had plans for him. We stopped to eat near Kingman, Arizona. Tim dropped me off just outside of Barstow, California where I slept in my tent just off Highway 58.

Chris McCandless Revisited

Four days ago (11 August), I was hitchhiking in Idaho and this guy picked me up. He told me that he went to school at Emory University in Atlanta, Georgia; he graduated in 1994. So I asked him about Chris McCandless (*Into the Wild* by Jon Krakauer) (McCandless went to school at Emory).

This guy said that he was three years behind McCandless in school. After McCandless' body was discovered in Alaska (1992), he was in an English class (in 1993?) with a professor that had taught McCandless a few years previous. The professor had the class study some of McCandless' papers.

This guy told the professor and the class that he thought McCandless showed a lot of hubris or suburban hubris when he tried to live in the wilderness of Alaska; he thought that McCandless was not well-prepared to live on his own. The professor and the rest of the class reacted very negatively to this guy when he used the word "hubris." This guy ended up getting a C- in the class.

A Hitchhiking Trip to Kansas

I got back this evening from a hitchhiking trip into Kansas. I am here at the shelter in Riverton [Wyoming]. I left Jackson on 26 October thinking that the Lord wanted me to go all the way to Washington, D.C.--but the Lord had different plans (which didn't break my heart); I really didn't want to hitchhike all the way to D.C. I made it all the way to the western edge of Topeka, Kansas and then I made my way back west.

The first day out of Jackson [Wyoming], I got all the way to Midwest where a couple let me stay for the night. He had spent 16 years in prison and we had a good talk about the things of God; he told me his girlfriend was schizophrenic--she asked me a lot of questions about the Gospel, but her head obviously was full of demons--I hope that my words were able to penetrate into her spirit. He drove a grader for an outfit in the Midwest area. I then hitchhiked to Gillette, then to Moorcroft and Sundance and made it to Lusk that night where I slept in a junked truck. The next day I hitchhiked to Valentine, Nebraska and then got a ride with a Christian truck driver to North Platte where he gave me forty bucks, so I was able to get a motel room that night.

The next morning I walked south out of North Platte and got a ride with a young Christian named Justin to McCook. He gave me a check for thirty bucks and I headed east to York where I slept in a grain box in an empty cattle shed that night. The next day I got to Salina, Kansas and then got a ride to Junction City, where I slept in a partially finished building that some construction company was still working on. The next morning around six o'clock I was awakened by the sheetrock crew: they weren't that surprised to see someone sleeping on their job-site. I then hitchhiked to Manhattan (where I tried to stay at the shelter, but it was full) and then to Topeka, Kansas. It was there that the Lord told me to head back west, so I cashed Justin's check in Topeka and then headed west on I-70.

This young man and his son picked me up outside of Topeka and took me to the Manhattan exit. He had been in the Army for six years and had spent some time in Iraq. Earlier this year he had taken a .44 magnum and tried to blow his head off, but failed. He was wearing sunglasses and there were some scars on his face and he was minus a few teeth. He said he should be dead, but that some higher power was looking over him. I gave him my CD hoping that some Scriptures might be en-grafted into him. I really didn't want to dig too deep into his life because I thought that he was going through a lot of stress from the Iraq War, but we did talk about the things of God for a while. He seemed like he was interested in my work on the road. I slept in a building off the interstate that night and then made my way west the next morning towards Junction City.

I was hitchhiking west out of Junction City when this young man picked me up and took me to his dad and mom's place in Enterprise, Kansas. His dad was involved in a local prison ministry; we had a real good chat. They let me stay with their family that night. We went to a Wednesday night service at an Assembly of God in Abilene, Kansas; it was excellent fellowship. The pastor let me give a little talk on what the Lord was doing in my life on the road. The Holy Ghost was very present in that fellowship that night. The next day I got dropped off in Abilene and visited the President Eisenhower Center for a little while and then headed north.

On Kansas Highway 18 heading west, I got a ride with a guy named Mike who gave me a ride close to Minneapolis, Kansas on U.S. Highway 81. Mike lived south of Leavenworth, Kansas about 15 miles and drove a truck for a living. We stopped and had a short prayer meeting. Mike gave me ten bucks and then I headed north.

I got some rides to Mankato, Kansas, walked west for a while and then found a junked pickup to sleep in. The next day I got a ride to Smith Center with a guy named Joe who gave me 80 bucks. I then hitchhiked to Phillipsburg and got a motel room that night. The next day I walked quite a bit--maybe over fifteen miles. I later learned that there is a prison just east of Norton, Kansas. I got a ride to Norton, got a hamburger and then hitchhiked to Oberlin where I got another motel room for the night. The next morning I headed north on U.S. Highway 83 out of Oberlin, and got a ride to McCook, Nebraska and then got a good ride all the way to Valentine. I hitchhiked west on U.S. 20 and slept in a junked van in Cody, Nebraska.

This morning I walked west a few miles and got a ride with a Christian all the way to Shoshoni, Wyoming. We had a real intense talk about all kinds of things pertaining to the Gospel. I wrote down some Christian websites that I thought he might like to look up. We stopped in Chadron, Nebraska for lunch. He was raised in Iowa and now lives in Nebraska with his wife and son. I should be heading back west to Jackson in two or three days.

Western Kansas

Just got dropped off here in Colby, Kansas. It is very windy outside. I got two rides from North Platte, Nebraska to Colby today. I slept in some big steel container outside of North Platte last night and stayed out of the wind. I was very grateful.

Yesterday I hitchhiked from Chadron, Nebraska to Sterling, Colorado. I walked out of Sterling for a mile or so and went to this gas station to fill up my water bottle. As I took my water bottle out of my bag, I heard this guy ask, "Where you going?"

I turned around and said, "Imperial, Nebraska."

He motioned for me to get inside his car.

We had a very lively conversation all the way to North Platte. I was going to take U.S. 6 to Imperial just to say that I had hitchhiked through Imperial (my great-grandfather had a ranch near Imperial over a hundred years ago). I had driven through Imperial a long time ago.

This guy was an animal science professor at the University of Hawaii near Honolulu. He was driving to Cozad, Nebraska to see his dad. I asked him if he had ever heard of Jay Lush who was an Animal Science professor at Iowa State University in Ames.

"Oh, yeah. I have definitely heard of Jay Lush," he said. "He was one of the pioneers in animal genetics back in the 30s, 40s and 50s. He was considered a wonder boy back at Iowa State."

I told him that I had rented a room in Ames from Jay Lush's widow for three months back in 1983. Her name was Adaline Lush. Mrs. Lush and I had pretty much nothing in common. I went to school at Iowa State for two or so months and then dropped out to work on a ranch in western Nebraska. I worked at the ranch for three days and then hitchhiked to Santa Cruz, California to see a poet friend and her daughter (I later wrote a short story about this hitchhiking trip: "High Plains Drifter"--it was later published by *Ethos* Magazine).

I never met Jay Lush; I believe he passed away in 1982. I became very good friends with Mrs. Lush's neighbors: John and Jette Foss.

John and Jette said that Jay Lush was a wonderful and warm human being--and he was very brilliant. Dr. Lush was later inducted into the National Academy of Sciences. Lush Auditorium on the Iowa State campus is named after him. I believe I had three classes in Lush Auditorium. John Foss was a biochemistry professor and Jette had a Ph.D. in genetics.

I told this guy that I hitchhiked from Washington state to western South Dakota back in 1987 with a medical doctor. He went to school at Iowa State and med school at the University of Iowa. He knew Mrs. Lush. It's a small world.

Meeting a Hermit in Montana

Earlier today I was walking out of West Yellowstone, Montana when this van pulled over to give me a ride. This guy looked like he was in his sixties or seventies. He had long, white hair and a short-cropped, white beard.

"I'm going to the junction of Highway 287," he said.

"Sounds good," I replied. "I'm heading south to western Wyoming."

"I live near Raynold's Pass. I'm a hermit."

"Yeah. I know where Raynold's Pass is. It's on the Idaho-Montana border heading to Ennis."

I asked him if he came from New York because I thought he had a New York accent. "I'm from Philly," he said.

As we drove down the road, he asked me, "How's hitchhiking? Is it easy to get rides?"

"Some days are fast and some days are slow. I get a ride when I need one. I'm on God's time."

The old man smiled and said, "You got that right." I told him that I was a Christian and he shook my hand.

"I know of a hermit that lives in eastern Montana. Are there any hermits in Iowa?" Earlier in the conversation I had told him that I was raised in Iowa.

"I don't know," I replied. "I haven't lived in Iowa for several years."

"So you are a hermit?" I asked. "Have you ever heard of 'The Three Hermits' by Leo Tolstoy?"

"Nope. Can't say that I have."

"It is a really good short story. If you want, you can read it on my blog." So I wrote down my blog address on a piece of paper and handed it to him.

I also told him that my book *High Plains Drifter* was at the West Yellowstone Public Library, if he wanted to read it.

The old man drove me to the intersection of U.S. 287 and U.S. 20 near Henry's Lake. I thanked him for the ride and got out of the van.

He smiled and drove off. I walked south on U.S. 20 till I got another ride.

*****

"You can preach a better sermon with your life than with your lips."

--Oliver Goldsmith

JUMP!

Back in 1996, I was hitchhiking in Utah and was dropped off in Evanston, Wyoming. There was a hitchhiker already there at the ramp sitting on the shoulder.

I walked up to him and asked, "How long've you been here?"

"I've been here since yesterday. Where you headed?" he asked.

"East. Probably Iowa."

"Have a good trip."

"Hope you get a ride soon."

I walked on down the ramp maybe fifty yards from where he was sitting and set down my bag. A while later, a third hitchhiker was dropped off at our ramp. He walked by me, said hello and walked to the interstate (I-80). He stopped on the shoulder and started thumbing.

Within half an hour, the first hitchhiker was picked up by someone and I waved at him as they drove on by. I was happy that he got a ride.

A little while later, I noticed a Highway Patrol car stop on the interstate where the third hitchhiker was standing. I could see the Highway Patrol talking with the hitchhiker. Within five or so minutes, that hitchhiker was swarmed by three or four more cop cars. I saw them put him in handcuffs, put him in the back seat of one of the cars and drive off.

A few minutes later, a Highway Patrol pulled up and stopped where I was standing. He checked my ID.

I asked him, "Why did they arrest the hitchhiker on the interstate?"

The Highway Patrol replied, "He had a warrant for his arrest in Idaho." He added, "Technically, it is illegal to hitchhike in Wyoming, but go ahead, it's all right with me."

"Thanks."

The Highway Patrol drove off.

Five minutes later, a Deputy Sheriff pulled up to me and asked, "Did the Highway Patrol check you out?"

I said, "Yeah."

"It is illegal to hitchhike in the state of Wyoming," he said in a stern voice.

"The Highway Patrol told me it was all right if I hitchhiked."

The Deputy Sheriff said something and then quickly drove off. He wasn't too pleased with what I had said.

So I stood there on the ramp a little nervous because I didn't want to be illegal.

A few minutes later, this pickup pulled up to me and the driver motioned for me to get inside. I got in the cab of the pickup grateful that someone was going to give me a ride.

The driver said, "I heard your name over the police scanner. They said that you weren't wanted for anything, so I knew you were all right."

"Thanks for stopping."

We drove on down the interstate. He told me that he was married and had five kids near Rawlins. He went to school at the University of Colorado in Boulder. He used to do a lot of rodeoing; he came across as a real cowboy.

"One time," he told me, "I was driving on I-80 here in Wyoming. I was driving an open-air Jeep. I picked up this hitchhiker. We were going down the highway and the hitchhiker pulls a knife on me."

I looked at him and said, "Whoa, Nelly."

"All that did was piss me off. I began to accelerate the Jeep. Now we are going 85 miles per hour. I keep my gun just to the left of my seat. I pull out my gun, point it at him and tell the idiot, 'JUMP!'"

I started laughing.

He continued, "The last I saw of him, there was a cloud of dust in the ditch."

Sometimes people mess with the wrong Marine. In this case, that hitchhiker messed with the wrong cowboy.

We drove to Rawlins where I met his wife and kids. He later drove me to Casper where he dropped me off on the ramp to I-25. I got another ride or two and was dropped off near Glenrock. I walked a short while, jumped over the fence and slept in the sagebrush that night.

Why is Hitchhiking Illegal in Wyoming?

LETTERS

Why is Hitchhiking Illegal in Wyoming?

I have hitchhiked through Jackson many times over the years. I have met some very friendly people as I have hitchhiked over Teton Pass, north to Moran Junction or south towards Pinedale. But why is hitchhiking illegal in Wyoming?

In September of 2009, I was camped out south of Riverton. The local police were driving by and noticed my tent by the river. They stopped and asked for my ID. They ran a check on my ID and told me that I had a bench warrant for my arrest. I asked them what the bench warrant was for. They said it was from an unpaid hitchhiking ticket. The fine was $60.00; I had 50 bucks on me.

They handcuffed me and drove me to the jail in Riverton. They asked me if I knew someone who could help out with the last 10 bucks of my fine. I gave them a phone number of a friend of mine. I spent a half hour in the Riverton holding pen till my friend arrived to help pay the rest of the fine.

I have no complaints with the Riverton police: they were friendly, professional and helpful. I can see why hitchhiking would be illegal on the Santa Monica Freeway in Los Angeles because it is so crowded with fast-moving traffic. But Wyoming? Wyoming is the most sparsely-populated state in the Union. Some people in various parts of this planet hitchhike just to get to work.

Is hitchhiking illegal because it is potentially dangerous? So isn't rock climbing, scuba diving and surfing. People get killed skiing, coal miners get killed, lumberjacks get killed. Most people die in bed. You're going to die eventually, so get used to it. Life is risk.

If anyone is interested in my life of hitchhiking (I have been hitchhiking for most of 14 years now), you can go to the Victor, Idaho public library and read my book *High Plains Drifter: A Hitchhiking Journey Across America.*

Support your local hitchhiker.

--Tim Shey, hitchhiker
JH Weekly
Jackson Hole, Wyoming
December 1-7, 2010
Volume 8, Issue 50

Outside the Box

Hebrews 13: 11-13: "For the bodies of those beasts, whose blood is brought into the sanctuary by the high priest for sin, are burned without the camp. Wherefore Jesus also, that he might sanctify the people with his own blood, suffered without the gate. Let us go forth therefore unto him without the camp, bearing his reproach."

"Let us go forth therefore unto him without the camp, bearing his reproach."

If a man is truly led by the Holy Ghost--a life of obedience to the Holy Ghost--then his life will be lived outside the camp--outside the box--bearing His reproach. Too many Christians put God in a box--which is idolatry. I like to tell people that I am like John the Baptist--I am living on the other side of the Jordan River. John the Baptist's living on the other side of the Jordan River was a powerful testimony against the Phariseeism that had infected the temple in Jerusalem. Phariseeism has also infected many Christian churches; I am sure that this has been a problem since the first century.

Phariseeism is a way to control people through the traditions of men. It rejects the leadings and spontaneity of the Holy Ghost. Phariseeism is a man-made religion that has a form, but no power and no life. Jesus said that He came to bring life and life in abundance. A lot of the time, the abundant life is lived outside the camp.

This is part of the reason that the Lord has had me hitchhike. Even if I did not have the ability to speak, my life would be a powerful sermon that glorifies God. Simply put: I am living for God and not for man; I am doing the will of my Father and not the will of myself. My Father works and I work. My food is to do the will of my Father who sent me. My Father's will is for me to live outside the confines and constraints of the box--the traditions of men and any other satanic construct.

If we follow Christ, we die daily. Our plans are in submission to His plans. Take up your cross and follow Me. Many times I have thought that all this hitchhiking is foolishness--it is definitely foolishness to the world. But the foolishness of God is wiser than the wisdom of the world. There was a book written about Francis of Assisi called "God's Fool". It is very good. Francis lived the life of a beggar, but his life was a powerful sermon that still influences Christians today. I believe it was Francis who once said, "Preach the Gospel and sometimes use words." I would rather walk the walk than talk the talk.

Walk the walk outside the camp.

Smuggled Over Teton Pass

This morning a friend dropped me off at the post office in Wilson, Wyoming. After I mailed a postcard, I walked to the edge of town and started thumbing for a ride.

I waited around ten minutes and this big delivery truck pulled into the entrance of the lane where I was standing.

The driver stuck his head out the window and asked, "Where you going?"

"Montana," I replied.

He drove the truck a little further into the lane, stopped and jumped out of the cab. He and this other guy rolled up the back door of the truck and I looked inside. They were hauling furniture.

"I'm going to Victor," the driver said.

"Sounds good."

So I pushed my backpack up onto this pallet in the back of the truck and crawled inside. They rolled the door down and locked it behind me. The trip from Wilson to Victor, Idaho is around 24 miles over Teton Pass.

As I rode in the darkness of the back of the truck, it felt like I was being smuggled. I think this was the first time I had gotten a ride in the back of a delivery truck. I could hear the roar of the engine as the truck labored up the mountain grade. A little later, I could feel the truck level out at the pass and then descend toward the Wyoming-Idaho Stateline. The truck picked up speed as the mountain grade leveled out and we approached the outskirts of Victor.

The truck stopped. A few moments later, the back door rolled back up. The morning light flooded the back of the delivery truck. I jumped out and grabbed my backpack. I looked around and noticed that they had stopped at the gas station on the south side of Victor.

I talked to the driver for a short while. Then the other guy appeared. I told them that I had been hitchhiking the States for a number of years.

Then I said, "You know, as I was standing in the back of your truck, I felt like an illegal alien being smuggled over the pass."

They both started laughing.

The passenger looked at the driver, smiled and exclaimed, “That is exactly what we were saying as we drove to Victor!”

The two guys who gave me the ride looked like they were of Hispanic origin. All kinds of Hispanic people are being smuggled over the southern border of the United States probably every day. How many Hispanics are smuggling white guys over Teton Pass?

After I shook their hands, I walked into the town of Victor and headed to a friend’s house feeling slightly illegal.

Reckless Faith

My Utmost for His Highest
By Oswald Chambers
June 18
"Don't Think Now, Take the Road"

"The wind was actually boisterous, the waves were actually high, but Peter did not see them at first. He did not reckon with them, he simply recognized his Lord and stepped out in recognition of Him, and walked on the water."

"If you debate for a second when God has spoken, it is all up. Never begin to say--'Well, I wonder if He did speak?' Be reckless immediately, fling it all out on Him. You do not know when His voice will come, but whenever the realization of God comes in the faintest way imaginable, recklessly abandon. It is only by abandon that you recognize Him. You will only realize His voice more clearly by recklessness."

Once in a while, especially in bad weather--rain, snow, bitter cold--people will ask me why I am hitchhiking (because to them it doesn't make any sense that I am hitchhiking in such inclement weather), so I tell them that the Lord told me to. It doesn't matter about the weather or what people think or say or what I want or don't want: the will of our heavenly Father cuts through everything in heaven and earth like a hot knife through butter. Sure it seems crazy what the Lord tells you to do, but we are called to be a peculiar people. I love the way Oswald Chambers says it: "recklessly abandon": recklessly walk across the water; recklessly hitchhike where people don't think you should hitchhike; recklessly obey the Lord and ask questions later. Live a recklessly abandoned obedient life or die. Live free or die.

Reckless teeth, reckless taste, reckless eyes, reckless everything.

If I am ever on a desert island, there are two books that I would have to have with me: the King James Bible and Oswald Chambers' *My Utmost for His Highest*.

Hitchhike recklessly or don't hitchhike.
Without reckless hitchhikers we would be a people no more.
Reckless obedience. Reckless faith.

Some Days Are Slow and Some Days Are Fast

I think it was in the spring of 2007 and I was hitchhiking from Washington, D.C. back to Wyoming. I got dropped off in Illinois on I-70 some place and camped there that night. The next morning I hitchhiked to East St. Louis.

I have heard horror stories of East St. Louis and that there are some real rough areas in East St. Louis. So I thumbed for a short while on this on-ramp trying to head west through St. Louis, Missouri. I then decided to walk across the bridge that spans the Mississippi River.

I walked across the bridge--it was the I-70/55 Bridge that goes into downtown St. Louis. The Mississippi River is very wide there in St. Louis. Looking at the Mississippi, you could see/feel the awesome power of the river as it headed south towards the Gulf of Mexico. I once had to walk across the bridge from Memphis into West Memphis, Arkansas. The Mississippi is even wider there than in St. Louis.

So I walked around downtown St. Louis for a short while and then tried to hitchhike on the on-ramp to I-70. No rides, so I moved on.

I would walk for a while and then try to hitchhike for a while. No rides. I did that three or four times and got tired of not getting any rides, so I just kept walking.

I had to walk from downtown St. Louis till I got into the suburbs--and it took all day. One thing that I noticed about St. Louis is that there are blocks and blocks of brick houses. I have never seen so many brick houses in a city before. I didn't walk on I-70 (the shoulder) through St. Louis, but I tried to walk parallel with it or tried to keep within a few blocks of I-70.

When I got into the suburbs, I began walking on the shoulder of I-70. I tried thumbing again as I walked on the shoulder, but nobody pulled over to give me a ride.

Then I heard this car slow down as it approached from behind me. When I turned to look at the car, the driver sped off. The same car did this three times in the span of half an hour. Strange.

I walked a while further and soon it was sundown. I walked into this industrial area and slept under this tractor-trailer that night.

The next morning I walked a short ways and got my first ride. I got two or three rides across the state of Missouri. I didn't have to walk very far between rides.

Then I got a ride to Kansas City, Kansas. I was dropped off on the west side of Kansas City. I walked maybe a quarter of a mile and this tractor-trailer pulled over. I hopped in and this guy gave me a ride all the way across Kansas to Denver.

Wow. What a fast day of hitchhiking. What a welcome relief from the day before. The Hand of Providence sometimes delays us for a reason and sometimes we get some place very fast. We could call it the Divine Timetable. Whenever I arrive some place, it always seems like perfect timing.

"In patience possess ye your souls."

Bereshith

Yesterday, I got a ride with a young couple from Israel. He was from Hebron and she was from Bethlehem. They gave me a ride for ten miles and then dropped me off in Last Chance, Idaho, the next town north of Ashton. They said that they were Orthodox Jews and that they needed to stop travelling for the day and prepare supper before Shabbat (the Sabbath). It was late Friday afternoon and the Jewish Sabbath begins on Friday evening and lasts till Saturday evening.

He said something very interesting. He told me that a rabbi (probably in Israel) warned that Japan was going to suffer a nuclear catastrophe. The rabbi said this before the earthquake in Japan in March of this year. The day before the earthquake, the government of Japan signed some sort of deal with the Palestinians. When you oppose Israel, bad things happen to you. He who blesses Israel shall be blessed, he who curses Israel shall be cursed.

I know very little Hebrew, but I believe that that Israeli couple were speaking Hebrew with each other. When he dropped me off at the gas station in Last Chance, he asked me if I knew the Hebrew word *bereshith*. I was a bit surprised because I DID know what *bereshith* means. *Bereshith* means "in the beginning" or "beginnings". What are the odds of an Orthodox Jew asking me what a Hebrew word means when it is one of the few Hebrew words that I know? A Providential ride.

He also said that he noticed while travelling in the United States that Americans for the most part are polite, but lack in hospitality. Whereas in Israel, the people are not very polite, but are very hospitable. He said that there are a higher percentage of hitchhikers in Israel than in the United States.

I Should Go To Dairy Queen More Often

This is from *The Mission* blog (Randy Sheets):

Please read the following, doctrinally correct, words by former Moody Memorial Church Pastor Harry A. Ironside (1876-1951). . .

"The Gospel is not a call to repentance, or to amendment of our ways, to make restitution for past sins, or to promise to do better in the future. These things are proper in their place, but they do not constitute the Gospel; for the Gospel is not good advice to be obeyed, it is good news to be believed. Do not make the mistake then of thinking that the Gospel is a call to duty or a call to reformation, a call to better your condition, to behave yourself in a more perfect way than you have been doing in the past. . .

"Nor is the Gospel a demand that you give up the world, that you give up your sins, that you break off bad habits, and try to cultivate good ones. You may do all these things, and yet never believe the Gospel and consequently never be saved at all."

SOURCE: Dr. Harry A. Ironside, from the sermon: "What Is The Gospel?"

Clearly, Ironside taught a Free Grace view of the Gospel, which is Biblical.

A changed life is the FRUIT of genuine repentance; and not a part of the ROOT of saving-faith.

--Traveller (Randy Sheets)

COMMENTS:

Amen.

Being raised in an idolatrous Irish Catholic family, I had sacraments running out of my ears. These sacraments are only man-made, outward shows of religion. The carnal, unsaved mind loves sacraments because it feeds their pride of self-salvation (work yourself for salvation).

We are saved from the inside out, not the outside in (by using sacraments). If someone is constantly concerned about external rituals and ordinances and liturgies, they can't possibly be abiding in Christ. Abiding in Christ is an internal relationship. External things like sacraments are only window dressing and are absolutely worthless.

It is only faith in the Blood of Jesus that cleanses us from sin--not external, religious calisthenics (sacraments).
--Tim Shey

This happened just a few days ago. I hitchhiked from North Bend to the west side of Eugene, Oregon. I had to walk with my backpack in 90 degree heat through Eugene and through Springfield (which is a number of miles). By the time I got to the east side of Springfield, my feet were aching and I was very tired.

I saw this evangelical church. They had a sign that said they were going to have a concert and free hot dogs. All I needed was to fill up my water bottle, because it was so hot. There were chairs set up outside the church and some people were setting up musical equipment. I put my backpack down next to the building and grabbed my water bottle.

I walked up to the people and asked if I could fill up my water bottle. The leader looked at me and hesitated (he probably saw my backpack) and then he said okay. I walked inside the church building and filled up my water bottle. I thanked them and tried to start a little conversation by mentioning that I was into intercessory prayer; there was no response.

I walked back to my backpack. Nobody invited me to their concert, nobody asked me if I needed a hot dog, nobody suggested that I sit down for a while and rest my tired feet (I was hobbling a little because I had been walking for miles in the heat with a 55 pound backpack). I walked away from that church building knowing that the spirit of Christ was not there--but I am sure they thought they were Christians because they went to a church building on Sunday to socialize (idolatry).

I walked on down the street and noticed a Dairy Queen sign. I had a few dollars on me, so I walked into the Dairy Queen. I bought my first cherry milkshake in years. After I finished my milkshake, I walked outside to my backpack. There was this family sitting at a table outside near my backpack. As I grabbed my backpack, they asked me what I was doing.

I told them about my life on the road and that I was obeying the Lord. We had the most wonderful fellowship. They wished me good travels as I walked on down the street. It was SO redeeming. That fellowship was so spontaneous and full of the Holy Ghost; they were genuinely interested in my life of hitchhiking and obeying the Lord.

That family at that Dairy Queen were so alive in Christ. That evangelical church was absolutely dead.

I should go to Dairy Queen more often.

--Tim Shey

Quote: "I should go to Dairy Queen more often." Tim, this speaks volumes and volumes beyond what we both realize my brother. I read your thoughts and it really touched me, I could visualize you standing there in the midst of a "church social club" and them not even recognizing some of the most basic tenants taught by Christ. They are unaware of Hebrews 13:2: "Be not forgetful to entertain strangers: for thereby some have entertained angels unawares;" and Matthew 25:35: "For I was an hungred, and ye gave me meat: I was thirsty, and ye gave me drink: I was a stranger, and ye took me in;" or Romans 12:13: "Contribute to the needs of the saints and seek to show hospitality." NOPE NONE OF THESE. . . So, where do you find solace and comfort and fellowship? In a Dairy Queen... Very very revealing and telling...

God bless you my brother.

--Randy Sheets

Saving faith is believing-in-God faith, not believing-in-a-man-made-institution faith or believing-in-myself faith. "Abraham believed God and it was imputed to him as righteousness."

I meet so many people who believe in Christian principles, but they reject Christ with their lives.

--Tim Shey

Years ago I was hitchhiking in West Virginia. I got dropped off in Charleston. I walked past this place where it looked like these people were having a picnic. I walked a couple of blocks away, put down my backpack and started thumbing for a ride.

A few minutes later these two teenage girls walked up to me and asked me if I would like to come to their church picnic. I was pleasantly surprised. So I walked back to their picnic. There I met the pastor and some other people. It started to rain, so we moved everything inside the church building. We had food and good fellowship.

I later learned that that pastor invited another hitchhiker and a homeless person to their picnic. Now how many pastors would do that? The pastor put me up in a motel for the night and I hitchhiked to Washington, D.C. the next day.

It was all so totally unexpected and spontaneous and refreshing. The Kingdom of Heaven is like a little child: unexpected, spontaneous, refreshing, living by faith in God.

--Tim Shey

North of Brookings

Today I hitchhiked from Trinidad, California to Brookings, Oregon. I walked a couple of miles north of Brookings on U.S. 101 to Lone Ranch Beach. My feet are fatigued from all the walking in the past two days. This is a beautiful, restful place to sit and listen to the Pacific waves crash on the beach.

For most of the past two months, I worked for some friends in Cedarville, California. They said that I could stay as long as I wanted. During that time I pruned a lot of trees on their properties, we fixed some fence, branded a few calves and I did a little painting on their house. I thought that maybe the Lord was going to let me settle down permanently there in the California Outback, but here I am on the road again.

It is pleasant, partly cloudy and cool--probably 60 plus degrees F. Some seagulls are flying around the beach. There is a small boat a mile from where I am sitting. Massive rocks stand like immovable sentinels on the beach and in the water. I am sitting on one of six picnic tables that are roughly a hundred yards from the water's edge.

The Son of Man Hath Not Where to Lay His Head

Luke 9: 58: "Foxes have holes, and birds of the air have nests; but the Son of man hath not where to lay his head."

The last two nights I slept in somebody's house south of Alpine, Wyoming. Three nights ago I slept on a couch at Jeremy and Felice's place. Before that I slept on a stack of lumber in Boise, Idaho; I slept in the back seat of a car at a repair shop in Ontario, Oregon; I slept in a trailer home converted into a warehouse north of Hamilton, Montana. Is transiency godliness? No, not necessarily, but in my life, transiency is obedience to the Lord.

Foxes have holes, birds have nests and I have the back seat of cars, the cabs of trucks, haystacks, corn stacks, under bridges, barns, abandoned homes and other assorted buildings to sleep in. The earth is the Lord's and the fullness thereof. The Lord definitely finds me some unusual and unique places to sleep.

Looks like I will head north into Montana in a couple of days. The Lord's thoughts are higher than my thoughts. As usual I will meet someone or several people and we will have a profound talk about the Gospel and/or the Lord will reveal something to me along the way.

Wyoming, Idaho and Montana

About a week ago, I hitchhiked from Jackson through Salmon, Idaho to just south of Darby, Montana where I slept in a barn. Walking out of Tetonia, Idaho, this guy picked me up--he had picked me up before in 2001 or 2002. I had given him my typescript *High Plains Drifter* when he first picked me up and he said he really enjoyed it. He gave it to his wife and she liked it, also. So when he dropped me off near Sugar City, I gave him a copy of my CD so he could read my other book.

After sleeping in that barn near Darby, I hitchhiked to Kooskia, Idaho and stayed with Kim and Pat. It was very hot that day—it might have been up to 100 degrees. It was very hot when I walked out of Mud Lake, Idaho the day before. I helped Kim and his son stack some lumber he bought for his house. His son gave me a 50-dollar gift credit card and I was very grateful.

The next day I hitchhiked out of Kim's place and ended up south of Whitebird, Idaho where I got a ride all the way south through Boise to Mountain Home. I slept outside in somebody's pasture about five miles east of Mountain Home. I didn't sleep well because of the mosquitoes.

The next morning I walked for a while and got a ride with a Christian who bought me some lunch. He dropped me off in Ketchum where he worked as a draftsman. I then hitchhiked through Stanley and Clayton. I walked a mile or two and slept outside on this mountainside. It was nice and cool and there were no mosquitoes, so I slept very well that night.

Victor, Idaho

Yesterday I hitchhiked from Dubois, Wyoming to Lander over South Pass to Farson and then to Jackson. From Jackson I got two rides over Teton Pass to Victor, Idaho. I camped out in my friend's back yard in Victor last night; I don't think it got down to freezing. It seems like it has been warmer than usual this September and early October in the Montana-Idaho-Wyoming neighborhood. I will camp out again tonight then head up into Montana tomorrow, God willing.

I hitchhiked from Riverton to Dubois two days ago and spent most of the day at the library there. I camped out near the river that night. If you are hitchhiking through Wyoming in the middle of the winter and you get dropped off in Dubois and don't have a warm place to sleep, you can sleep at the Episcopal Church--I have slept there at least four times now.

From mid-August till late September, I hitchhiked in Washington, Oregon and California. I helped some friends in Dayton, Washington and did some work for some friends in Cedarville, California. I hitchhiked between Dayton and Cedarville a few times during that time. Eastern Oregon is very beautiful on U.S. 395. I passed through Pendleton, Oregon during the big rodeo there--it was the 100th Anniversary of the Pendleton Roundup--there were tons of people there.

On my way through Idaho on U.S. 12 to Montana, I lost my sleeping bag (I had two summer sleeping bags). I hitchhiked to Helena, Montana, and with the money I made working for my friends in Cedarville, I bought a brand, new sleeping bag. It is rated at 10 degrees F--so it is the warmest bag that I have ever had. It has been a cooler summer this year than last year: could this mean that it will be a colder winter than last year? I am grateful to have a warmer sleeping bag now that we are getting closer to winter.

Elvis

I just got back from a short trip up to Kooskia, Idaho. I left on Monday morning (13 November) and hitchhiked all the way to Missoula, Montana where I slept in a junked RV. The next day I got to Kim and Pat's place near Kooskia and stayed one night. The next morning Kim and his son-in-law, Blake, took me to Grangeville and it took me two more rides to get to New Meadows, Idaho. I got a hamburger and then checked my e-mail at the local library. I then walked maybe two blocks and got a ride to Boise and then to Mountain Home with a guy named Elvis.

Elvis was around forty years old and was driving a tow truck; he was a Christian and we had a real intense talk about the things of God. I look back on the past five days of hitchhiking and my talk with Elvis was very significant. The Lord has done some very power things in his life. He was in and out of prisons for twenty-two years; he had a conversion experience while in prison. Since he has been out of prison, the Lord has blessed him with a good-paying job, a home and a vehicle (or vehicles). He is now raising his eleven-year-old daughter by himself. Elvis had a lot of good insights into the Gospel. It is really powerful how the Lord really turned his life around. Elvis lives by faith—and it is through faith that we are able to please God. Elvis and myself stopped by his house in Boise where I met his daughter, then he drove me to Mountain Home and got me a motel room.

After Mountain Home, I walked quite a ways and got a ride with a guy who lived in rural Bellevue, Idaho. We stopped by his house where I met his girlfriend. She made me a sandwich and I was able to dry my shoes and socks out while we talked. I then hitchhiked to Arco where the sheriff's department put me up in a motel room for the night. I was trying to find a junked vehicle or a building to sleep in, but this guy saw me walking around town and stopped his van behind me and told me to go to the sheriff's department and that they would help me out. I was grateful: it got down to fifteen degrees last night.

This morning I hitchhiked to Idaho Falls and then to Sugar City and finally made it to Jackson this afternoon around 4:30. I believe I could be heading east in a couple of days. My meeting with Elvis was very important; it was excellent fellowship.

Marty the Stonemason

A week ago I hitchhiked out of Jackson and this guy picked me up. I listened to him talk for a few minutes and then I said that he had picked me up before—three years ago outside of Dillon, Montana. He looked at me and said something like, "Hey, you're that hitchhiker I picked up." He then repeated what I had said three years ago—something about being led by the Holy Spirit. I remember when I met him up in Montana it seemed that he thought it was very peculiar that I was "obeying the Lord" and "being led by the Holy Ghost."

His name was Marty and he worked as a stonemason in the Jackson area. He told me that when I first met him he was drinking a lot and doing drugs and that his life was a mess. Then he said that three months ago he became born again—he even showed me his Bible. I was pretty happy for him and I shook his hand. He dropped me off between the Wilson Bridge and Wilson and I proceeded to walk up the road toward Teton Pass.

I got a couple of good rides to Three Forks, and then to Missoula, Montana. I got a short ride to Lolo and slept in a shed that had some plywood in it. It rained that night, so I was grateful to be high and dry.

The next day I hitchhiked to Kooskia and stayed at Kim and Pat's place for one night. I helped Kim screw down his deck and then we stained it later that day. The next day I hitchhiked to Boise and stayed in a motel. When I hitchhiked from Lolo to Kooskia, this guy named Kelly picked me up and kicked me down 170 dollars. He was a Christian and we had great fellowship. Kelly drove me all the way to Kim and Pat's place. So I used the money that Kelly gave me and got a motel room in Boise. I was very grateful because I was so tired.

The next morning I left Boise and got to Mountain Home. As I walked east on U.S. Highway 20, I noticed a wallet in the ditch. I was curious and picked up the wallet—there were some credit cards lying on the ground. I looked at the driver's license and it expired in 2011. So I gathered up the credit cards, put them back in the wallet and put it in my backpack. I wanted to send the wallet back to its rightful owner (I later mailed it from Clayton, Idaho).

I got rides through Fairfield, Hailey, Sun Valley and Stanley. I got dropped off just west of Yankee Fork where I slept in the woods that night. There was a lot of smoke in the air because of a forest fire in the area.

Cal

Today I hitchhiked from Billings to Bozeman. Just outside Laurel, this guy named Cal picked me up and gave me a ride to Livingston. He was going all the way to Missoula. We had a good conversation.

I made the comment, "Most people will probably think that since I am a hitchhiker that I am living on the fringes of society." Cal said, "No. You're not living on the fringes. Most of the people out there live on the fringes." I think I know what he meant by this. Most people are so conformed to the world system that they aren't really living—they only exist. My life is very rich: I'm growing stronger in my faith in God; I have met some great people on the road and I have seen a lot of beautiful territory throughout the United States.

I told Cal that he should write a book about his life. Cal told me that he was 51 years old, had been in the Navy SEALs for several years, spent a couple of years at Harvard and was involved in the banking industry. He had been married and had some kids and wanted to move to Bolivia. Cal had a brain injury back in 2001 and was slowly recovering. I told him that he had experienced quite a bit for one life.

Truth in the Inward Parts

Psalm 51: 6: "Behold, thou desirest truth in the inward parts: and in the hidden parts thou shalt make me to know wisdom."

The Lord doesn't want the truth in your head or on your tongue (lip service)--He wants truth in the inner man. If truth is in your heart, then people will eventually hear it come out of your mouth; they will see it in your life.

The Lord is very subtle. Sometimes, when He speaks to you, you don't catch it the first time. But if your heart is right with God, He will most definitely make His will known to you in various ways. It must bear witness with your spirit, "the inward parts"--"the hidden parts." He will use circumstances to get your attention, but it all must line up with the inner man. Sometimes this is my biggest complaint with the Lord: He is so subtle. Time and again I say, "You are too subtle, Lord!" He will keep repeating Himself, and if I don't get it, He will chasten me. He chastens those whom He loves--that's how we learn obedience. If the Lord does not chasten you, you are a bastard and not a son. Of course, I perceive the subtleties of the Lord much better than I did twenty years ago. The more we abide and learn of the Lord, the more He develops our inner man.

I suppose Satan can counterfeit this. Satan tries to counterfeit anything the Lord does. If you believe you are hearing from God, and there is confusion in your life, then there is a good chance that you are not hearing from God. With the Lord there is a peace that surpasses all understanding. Of course, if you are young in the Lord and your inner man is not very well developed, then you are not going to hear from God. If you are living in sin, you are not going to hear from God. If you are in sin, Satan can speak to you in circumstances, dreams and visions--Satan can probably give you certain Scriptures to keep leading you away from God. Satan can twist and turn Scriptures to fit his own agenda.

There are so many benefits and checks and balances when we are seeking a deeper, more intimate relationship with Christ. I was hitchhiking somewhere near Manhattan, Kansas a few years ago and this young lady picked me up. She said, "The Lord told me to pick you up." I said, "Praise the Lord. I'm a believer, too." She then told me that two years before she was driving down the road and saw this older man hitchhiking, so in the goodness of her heart she wanted to stop and give him a ride. But the Lord said, No! So she drove on by. Two nights later, that same man was on the local nightly news and he had robbed and killed an older couple in the next town. It really pays to abide in Christ. If she had been a nominal Christian, maybe she would not have heard the Lord's voice telling her not to pick up that hitchhiker.

Truth in the inward parts, wisdom in the hidden parts: sounds like the Holy of Holies. Satan can attack our outer man, our mind (brain), our flesh, but Satan cannot penetrate the Holy of Holies of a man who is really abiding in Christ. There is a wall of fire surrounding him, protecting him. When a man is dead in his trespasses and sins, his Holy of Holies is dead and can possibly be demon possessed. When a man is born again, his Holy of Holies is revived and cannot become demon possessed. His Inner Court and Outer Court can be demon oppressed, but not his Holy of Holies. If a man turns his back on Christ and lives a reprobate life of sin (has lost his salvation), then I do believe that his Holy of Holies can become demon possessed. When a man is born again, Christ invades the Holy of Holies, and if we continue to abide in Christ (the full counsel of God), Christ invades and revives and redeems the Inner Court (the mind, the soul) and the Outer Court (the body). The Gospel of Jesus Christ isn't just for the spirit--it is for the body, soul and spirit.

I Thessalonians 5: 23: "And the very God of peace sanctify you wholly; and I pray God your whole spirit and soul and body be preserved blameless unto the coming of our Lord Jesus Christ."

A Bank Robbery

I was walking north of Sugar City, Idaho on U.S. 20 and this deputy sheriff stopped me and told me to sit on my backpack. He then called for some backup and, within a few minutes, there were two police cars and a sheriff's department pickup. The deputy sheriff told me that a masked man wearing a blue coat had just robbed a bank in Rexburg; I was wearing a blue coat.

So they frisked me for weapons and two guys showed up with body armor and Kevlar helmets—one of them carried an M-16. They went through my backpack and then they let me go. I have been frisked for weapons before, but not with a guy in body armor, helmet and an M-16. I hope they catch the guy who robbed the bank.

Shiloh

"Shiloh" was written in the late fall of 1996. I took a bus from Iowa to Texas and then to Las Cruces, New Mexico. I then hitchhiked north out of Cruces up through the Texas Panhandle and back to Iowa. I wrote "Shiloh" in three days when I got back from my bus/hitchhiking trip. It was first published by *Ethos* Magazine in 1997.

Shiloh
By Tim Shey
Brutal deathdance;
My eyes weep blood.
Pharisees smile like vipers,
They laugh and mock their venom:
Blind snakes leading
The deaf and dumb multitude.
Where are my friends?
The landscape is dry and desolate.
They have stretched my shredded body
On this humiliating tree.
The hands that healed
And the feet that brought good news
They have pierced
With their fierce hatred.
The man-made whip
That opened up my back
Preaches from a proper pulpit.
They sit in comfort:
That vacant-eyed congregation.
The respected, demon-possessed reverend
Forks his tongue
Scratching itchy ears
While Cain bludgeons
Abel into silence.

My flesh in tattered pieces
Clots red and cold and sticks
To the rough-hewn timber
That props up my limp, vertical carcase
Between heaven and earth.
My life drips and puddles
Below my feet,
As I gaze down dizzily
On merciless eyes and dagger teeth.
The chapter-and-versed wolves
Jeer and taunt me.
Their sheepwool clothing
Is stained black with the furious violence
Of their heart of stone.
They worship me in lip service,
But I confess,
I never knew them
(Though they are my creation).
My tongue tastes like ashes:
It sticks to the roof of my mouth.
I am so thirsty.
This famine is too much for me.
The bulls of Bashan have bled me white.
Papa, into your hands
I commend my Spirit.

Ethos
February/March 1997
Iowa State University

Genesis 49: 10: "The scepter shall not depart from Judah, nor a lawgiver from between his feet until Shiloh come; and unto him shall the gathering of the people be."

A Sleeping Bag in the Ditch

This morning I walked a mile west on I-90 outside of Belgrade, Montana. I tried thumbing to Three Forks, but there were no rides, so I thought I would walk back to the library in Belgrade and catch up on some reading.

I started walking back east when this highway patrol pulled over and asked, "Did you lose a sleeping bag in the ditch?"

"No, I sure didn't," I replied.

"It was around mile marker 295. Someone called it in. They thought maybe someone was sleeping in the ditch."

"No. I didn't get that far down the road. I'm heading back into town."

"All right. Take care."

If anyone was sleeping in the ditch last night, they would have to be pretty tough or have a very good sleeping bag. It got down to 2 degrees F last night.

I hitchhiked up from Victor, Idaho yesterday. I'll spend some time here at the library today. I have a friend here in town, so if he is home, I'll probably sleep on his couch for the night. If not, I'll just hunker down in my favorite junked pickup on the west side of town and sleep there. It is supposed to get down to -5 degrees F tonight. I slept in the pickup last night and stayed warm.

I was hanging out at the McDonald's last evening reading my Bible and drinking my Tea-Sprite Non Alcoholic Cocktail when I overheard this kid say that the Lewis and Clark Expedition (1804-1806) spent more time camped out near Three Forks, Montana than at any other place in their travels. There are three rivers that converge at Three Forks: the Jefferson, the Madison and the Gallatin. These three rivers form the headwaters of the Missouri River that meanders all the way to the Mississippi River north of St. Louis, Missouri. These rivers were named after President Thomas Jefferson, Secretary of State James Madison and Secretary of Treasury Albert Gallatin.

Eastern Wyoming

Yesterday was a good day for hitchhiking. I got two rides from Helena, Montana to Orin Junction, Wyoming. I had stayed at the *God's Love Mission* in downtown Helena for three nights and then I hit the road.

I walked two or three miles towards East Helena when this car pulled over to pick me up. This guy just got out of law school at the University of Minnesota. He was going to Crow Agency, Montana for a job interview. We had a good talk about Indian affairs and about the things of God. We stopped in Billings were he gassed up his car and bought me a meal at Burger King. He dropped me off in Hardin at a gas station.

I walked from the gas station to the interstate and just started walking down I-90 when this pickup pulled over. This guy was a pastor of three churches in Helena, Boulder and Townsend. We had a great conversation about the things of God all the way through Buffalo and Casper, Wyoming. We stopped at a Taco Bell in Casper and got something to eat. He dropped me off in Orin Junction (just south of Douglas, Wyoming) last evening at the gas station. We had a short prayer meeting and then he gave me a sandwich that he had with him. He was going to a funeral in Denver.

I slept underneath the bridge on U.S. 20 just east of Orin Junction last night. I slept there back in October of 2000. It wasn't too cold last night--maybe in the upper 20s. This morning I walked back to the gas station, got a cup of hot chocolate and then sat down in the trucker's lounge and watched The History Channel for an hour. There was a program on the weapons used by the Germans in World War II. I enjoy military history.

Just got dropped off here in Lusk, Wyoming. After I leave the library I will probably head east on U.S. 20 to Chadron and then to Valentine, Nebraska. Right now it is very foggy; there is a winter storm coming in later today. There might be accumulations of 2 to 4 inches of snow in this part of Wyoming.

Egypt is Burning

Yesterday I hitchhiked from Jackson, Wyoming to Bozeman to Columbus, Montana. I got some fast rides; the Presence of God was very strong all day. Last night I slept on a stack of lumber at the Timberweld place in Columbus. While I was laying in my sleeping bag, I began to compose "Egypt is Burning" in my mind. This morning I finished composing the poem at the McDonald's here in Columbus.

*****

Egypt is Burning
By Tim Shey

Sons of Ishmael,
The Scriptures have come full circle.
The angel of the Lord said
He would be a wild man.
Abraham's firstborn was Isaac.
Mount Moriah pointed towards Calvary.
Malachi said:
Was not Esau Jacob's brother?
The Lord said:
Jacob I loved, Esau I hated.
Cain murdered Abel;
Joseph was hated by his brothers.
Jesus was killed
In the house of his friends.
Hagar's offspring mocks
The Messiah to this day.
Egypt is burning.
Isaiah walks naked among you.
Your sin and rebellion is
Broadcast twenty-four seven

On *FOX* and *CNN*.
Israel is no longer Jacob:
He has power
With God and men.
Who can resist God's will?
The Lord is transforming
The bloody Middle East.
Shiloh is here in power:
He couches as an old lion.
The Tribe of Judah
Rules in Zion.
The City of David
Is a state of rest:
The Book of Hebrews, Chapter Four.
Those who abide in Him
Are already in New Jeru-Salem.
All you have to do
Is meditate on Genesis 49: 10.
Who is this
That cometh from Edom?
His Cross is splattered in red.
Egypt is burning.
I will tread them
In mine anger.
Egypt is burning.
The handmaid despised Sarai.
Egypt is burning.
Do not reject
His Precious Blood.
Egypt is burning.

A Walk in the Sun

Several years ago, I was walking on I-90 south of Hardin, Montana. I got a ride to Lodgegrass on the Crow Indian Reservation. Then I got a ride in the back of this pickup with some Crow teenagers. They drove me south a few miles and then dropped me off. They turned the pickup off the interstate, drove through the median and headed back north.

So I started walking. It was hot and sunny. I believe it later got up to 95 degrees that day. I ran out of water. I walked a few more miles and I was getting thirsty.

As I walked, I would scan the pasture land in that part of south central Montana for water tanks (water for livestock). I remember I ran out of water somewhere in New Mexico once and I saw this windmill about a quarter of a mile from the road. I jumped over the fence, walked to the windmill and it was pumping water into a tank. I filled up my water bottle and walked back to the road.

I noticed these trees growing by this fence. I walked into the ditch to take a closer look and I heard this whimpering sound. I looked in these bushes and there was this dog laying on the ground acting all scared and submissive. It looked like it was afraid that I might kill it or something. I knelt down and patted it and scratched its neck to reassure the dog that I wasn't going to harm it in any way.

I walked out of the ditch and back onto the interstate and the dog followed.

The dog and I walked a little further and I spotted this water tank in this pasture. I jumped the fence, walked to the tank and filled up my water bottle. I lifted the dog over the tank's edge and let the dog drink it's fill.

By this time we were past the Wyola turnoff. It was hot and I was concerned about the dog and myself having enough water.

People would drive past and honk and wave at us and I would wave back. I guess you don't see too many people walking on the interstate with a dog at their heels.

We ran out of water and we had walked a number of miles. Soon the sun went down over the western horizon.

About ten miles from the Ranchester, Wyoming exit, I saw a bag of potato chips in the ditch. I walked into the ditch, opened up the bag and had something to eat.

Around midnight I finally made it to the Ranchester exit. There was this guy sitting at the exit--he was wearing a cowboy hat--and it looked like he was waiting for a ride. I walked up to him and we started talking.

He told me that he was a truck driver and that his dad had died earlier that day. A relative phoned him about his dad's death, so he parked his tractor-trailer somewhere at a truck stop in Montana. He was going to hitch a ride back to his home in Florida.

"I walked a number of miles after my last ride," he said. "I threw out this bag of potato chips, because I really didn't need it."

"Thanks," I replied. "I ate the bag of potato chips. I saw them in the ditch. I guess I needed the extra calories."

The truck driver laughed.

He said, "Let's walk into town and I'll buy something for you to eat."

"Sounds great." I was grateful.

He looked at the dog and said, "You have a nice dog."

"It's not mine. I saw it hiding in some bushes and it started following me."

The truck driver knelt down and felt the belly of the dog.

"She's pregnant," he said. "Crow Indians like to eat dogs. Maybe you saved it from being killed and eaten."

We walked to Ranchester and into this bar. The truck driver ordered a pizza for me and I had a couple of Cokes. We stayed at the bar for half an hour and then walked to this park.

I had a sleeping bag and he didn't, so I gave him a sweater to help him keep warm. I crawled into my sleeping bag. These two girls had followed us from the bar in their car. They drove to the park and started talking with the truck driver who was sitting on a picnic table. They were drunk and very loud and they said a few things that I am not allowed to repeat here.

The truck driver told them that he needed a sleeping bag. The two girls drove off and soon returned with a sleeping bag. The two girls talked with the truck driver for a few more minutes and then drove off.

The next morning, I parted ways with the truck driver. I walked south of Ranchester on I-90 minus the dog. I guess the dog found a place in Ranchester for her to have her pups. And the Lord used me to lead her to her new home.

The Helena Hobo

A few days ago I was looking at the page that advertises my book *High Plains Drifter* on *Amazon.com*. I noticed a comment of a conversation I had with someone I met on the road. He had picked me up hitchhiking just south of Townsend, Montana on 17 December 2009. Here is the comment:

The Helena Hobo

"Me and a friend picked up Tim outside Townsend, MT on the way to Bozeman, MT. We gave him a ride to Three Forks. He was a kind, gentle man, with a belief in God that will never be shaken. As an agnostic, I was moved the way this man blindly follows the Lords words throughout the country."

Even though I have had very little experience riding freight trains (I have ridden two freight trains in my life), it was an honor that he called me a "hobo." I would like to make one little correction: I am not "blindly" following the Lord. The Lord inspires me (the Lord speaks to my spirit or the Lord's voice is perceived by my spirit) and in faith I follow Him on the road.

"Those who are led by the Spirit shall be called sons."

A Ransom For Many

Mark 10: 42-45: "But Jesus called them to him, and saith unto them, ye know that they which are accounted to rule over the Gentiles exercise lordship over them; and their great ones exercise authority upon them. But so shall it not be among you: but whosoever will be great among you, shall be your minister: And whosoever of you will be the chiefest, shall be the servant of all. For even the Son of man came not to be ministered unto, but to minister, and to give his life a ransom for many."

If Jesus gave His life as a ransom for many people, shouldn't we do the same? Time and again I think that walking all over the place with a backpack and hitchhiking and never staying in one place for very long seems so foolish. How long do I have to do this? When can I settle down and have a basic job and have a place to live? If being a ransom for many means that the Lord wants me to travel the highways and the byways and follow Him, then there is nothing I can do about it. I have been bought with a price: salvation is not cheap, grace is not cheap: it was the price of the Son of God when He was flogged half to death and then died on the Cross--with nails pounded through his hands and feet. We don't have to understand all that the Lord tells us. Yes, the Lord gives me understanding in some things, not all things. If we understood all things, then where would be the humility, the living by faith and revelation? Prideful people don't need to live by faith--they live by self-reliance, not God-reliance. The humble man has a broken and contrite heart and is seeking God daily: he obeys the Lord even though he does not understand what the outcome will be: he is living by faith and not by sight.

It looks like the Lord wants me to hitchhike west towards Bozeman and then back to Jackson. In my flesh I ask, why does the Lord want me to hitchhike so much through Bozeman and Jackson? I really get sick of it. It is so repetitious. At least I have a healthy hatred for sin and for the world system. Maybe that is God's ulterior motive: I really do hate the world with a perfect hatred. The love of the world is enmity with the Father.

A Peculiar Path

Ephesians 2: 10: "For we are his workmanship, created in Christ Jesus unto good works, which God hath before ordained that we should walk in them."

We are his workmanship. We cannot effect change in ourselves or in the world around us by our self-willed "good works". We abide in Christ and He leads us into the good works that He has already prepared and ordained for us. A good work is a God work--a Father's will work. There are many, many works in the earth, but most of these are self-will works, works of the flesh, works of pure reason. A good work is something that the Lord wants us to do, not what we want to do. God is always glorified in a good work; He is never glorified in a self-will work.

When I finished my college degree in May of 1995, I had no idea that the Lord would have me hitchhike around the United States so much. This is just a wild guess, but I believe I have hitchhiked coast-to-coast at least forty times (total hitchhiking mileage). If this is what the Lord wants, then so be it.

I look back to 1996 (when I began my hitchhiking after a nine-year hiatus) and I can see the Lord's workmanship in my life. So many things have died out in my life, so that the life of Christ can be exalted in me. Curses have been broken in my life: these things (demons) can only come out through prayer and fasting. I am physically, mentally and spiritually stronger because of these past nine years of hitchhiking and obeying the Lord. It is the power of that revolutionary act--the Resurrection--that has made me into the New Man.

"That we should walk in them." I know I have done a lot of hitchhiking, but I am curious how many miles I have walked. I bet I average seven to twelve miles per day. But then it is not necessarily the physical act of walking, but walking in the Light as He is in the Light. He lightens our path, our unique path--a path-with-a-purpose. Every man and woman is special and specially designed to walk that unique path that God has ordained. It may be a peculiar path. It may not bring fame, wealth or notoriety, but it will glorify Him and his workmanship. And whatever He works into us shall never go void.

A Great Multitude Followed Him

Mark 3: 7-8: "But Jesus withdrew himself with his disciples to the sea: and a great multitude from Galilee followed him, and from Judaea, and from Jerusalem, and from Idumaea, and from beyond Jordan; and they about Tyre and Sidon, a great multitude, when they had heard what great things he did, came unto him." These two verses are very interesting. It says a great multitude followed Him—from Israel and outside of Israel. How many Scribes and Pharisees were among this multitude? How many theologians? How many of those who indulge in the knowledge of good and evil? The common man could see that Jesus was special. Some came to the conclusion right away that He was the Messiah. Some were curious and eventually called Him a teacher or a prophet.

Where were the temple types, the box makers, and the people in authority who told people what to do but did not do it themselves? I believe the word "hypocrite" means actor. Don't act it; be it, live it. But as we know, the carnal, unsaved mind loves a power trip. The people in power—the temple priests—absolutely resented the life and work of Jesus. He was the real deal. He preached and taught with authority. He did not have a Ph.D. from Hebrew University. He lived by faith and revelation: His university was constant communion with His heavenly Father. Don't get me wrong: there are a lot of inspired books that will feed your spirit, but we must be connected to the Source of all wisdom first and then indulge in the great books second. First things first. No doubt Hebrew University and Moody Bible Institute and Iowa State University have some very learned people who are totally dedicated to God. But without a relationship with the Lord Jesus Christ, there is no wisdom—just dead knowledge, information, hot air and glorified nonsense. With the fear of the Lord is the beginning of wisdom.

This multitude that followed Jesus were hungry for something—bread from heaven. Jesus' words were spirit and they were life. Jesus Himself was bread that fed men's souls. Why is it that from the age of fifteen through the age of eighteen I wanted to commit suicide? I had plenty of food, shelter and clothing. I was an atheist at that time and my life was exceedingly miserable because my spiritual life was dead. I needed the bread from heaven to sustain me. I began to pray to God when I was eighteen and I could feel grace come into my life. I was twenty-two when I became born-again and then I became engrafted into the Tree of Life. Even though I struggled with various sins in my early Christian walk, I at least knew where my food came from and, if I prayed for wisdom, He would give it to me.

Jesus withdrew himself to the sea, yet the multitude still followed Him. I would rather be a poor man in America than a rich man anywhere else. Why? Because a poor man in Christ still has the freedom to worship God in America than a rich man in China or Europe. Technically, I am poor, but in Christ, I am very rich. In America I still have the freedom to follow Jesus. Why are so many people still trying to emigrate or illegally immigrate to America? Because of the Gospel of Jesus Christ. Without the powerful influence of the Gospel in this nation, America would be just another third-world country. The economy, the social mores, the structure of our government is built on the Rock of the Lord Jesus Christ. An unbeliever who immigrates to this country may say that he came here for a good-paying job. How did that good-paying job come about? From thin air? No. There is a freedom in this country that lets believers and unbelievers be very creative and inventive—which help to create jobs. As a hitchhiker that lives by faith, I bet I live better than 90% of the people on this planet. Why? Because I am with the multitude that follow Jesus. The Lord feeds me, He heals me, He casts demons out of my body, He gives me wisdom—He is my economy, my social mores and my government.

And from beyond Jordan. Looks like some of John the Baptist's buddies are following Jesus—you know, the drug addicts and prostitutes and prophets and hitchhikers who really don't fit into the satanic-temple-priest-boxmaker-go-to-church-on-Sunday-lip-service-Christian construct. These people that come from beyond the Jordan want life and life in abundance. This life—this Tree of Life—

does not come from the study of theology (theology means the study of God—God doesn't wants us to study and analyze Him, He wants us to abide in Him in faith—not through dead reason) or by jumping through man-made hoops spawned by the traditions and rituals of men. This life can come only through a childlike heart that yearns and wants to follow the Master wherever He leads us. Come and follow Me--not come and think about Me or come and analyze Me. I know that there are strong Christians who are considered theologians, but I must speak bluntly: theology by its very nature and definition is Satanic: it is the Tree of the Knowledge of Good and Evil. Leo Tolstoy once said that theology is the Satanism of religion. This tree has a little bit of good, a lot of evil and leads to spiritual death. It is a stench that rises up to heaven; it is a tree that bears only dead fruit. Before you can study a frog, you have to kill it first.

When they heard what great things he did. Look at their reaction to the work of Jesus. They ran after Him. Look at the reaction of the Pharisees: they wanted to kill Him. They eventually did kill Him and they are still trying to kill Him. A Christian Pharisee is probably the most wicked man on this earth. They are so sure of their doctrine, they are very self-righteous and they prevent others from entering into the Kingdom of Heaven. "You call yourself a prayer walker? You hitchhike because the Lord told you to hitchhike? You don't preach salvation? Where in Scripture does it say that?" The letter kills, but the Spirit gives life. Scripture interpreted by dead, Pharisaical, spiritless reason can only kill; Scripture interpreted by a childlike, Spirit-led heart can only give life. Some people miss Christ for the Bible verses: John 5: 39-40: "Search the scriptures; for in them ye think ye have eternal life: and they are they which testify of me. And ye will not come to me, that ye might have life."

Do we follow Jesus, or do we follow man? Do we follow revelation or uninspired reason?

And a great multitude from Galilee followed Him. A great multitude. Followed Him. They did not study or analyze Him--they followed Him.

The Spirit Driveth Him into the Wilderness

Mark 1: 12: "And immediately the spirit driveth him into the wilderness." This verse reminds me so much of what happened in my life back in 1986. I had been working on an apple farm near Embudo, New Mexico for two months. The people I was working for were unbelievers and it was fairly oppressive being around them. Then early one morning, the Lord told me to hit the road—the Spirit drove me from the apple farm to the highway and I hitchhiked all the way to Ellensburg, Washington.

It was all such a relief to get away from those dead, pagan people. I tried to talk to them about the Gospel, but it was like talking to a brick wall. The Lord blessed me so much for leaving that place. I eventually got into a small Pentecostal church in Ellensburg and it was a very good experience; I later got baptized in water by the pastor of that church—Pastor Coussart. That was the first Holy Ghost church I ever attended. It was called Bethel Gospel Church.

When the Holy Ghost drives you into the wilderness, or onto the highway, or into the city—let me tell you: you got to go NOW! Timing is everything. If the Lord wants you to go someplace immediately, then there is a good reason. It may not be your reason, but it will be His reason and that's all you need to know.

Why did the spirit drive Jesus into the wilderness? Because Jesus obeyed His Father, came to John the Baptist and asked John to baptize Him in the Jordan River. Then when Jesus came up out of the water a voice from heaven said, "Thou art my beloved Son, in whom I am well pleased." All these things happened because Jesus obeyed His Father.

If you don't obey the Father, then the spirit won't drive you anywhere. You will go nowhere fast. You will go backwards and trample on His Precious Blood: Hebrews 10: 38-39: "Now the just shall live by faith: but if any man draw back, my soul shall have no pleasure in him. But we are not of them who draw back unto perdition; but of them that believe to the saving of the soul."

Mark 1: 22: "And they were astonished at his doctrine: for he taught them as one that had authority, and not as the scribes." No doubt the Presence of God was so powerful in Jesus (the anointing) that His doctrine ran over their karma, which ran over their dogma. So many cars, so many dogs—so little time. When you abide in the Presence of God, it can really discombobulate (I think that is a word) some people. When you live and drive outside the box, you can only run over the karmas and dogmas that love to stay inside the box. The Scribes were inside the box. The Pharisees are inside the box. The box makers have built mansions—or tombs—with whitewashed walls inside the box. So many boxes, so little time.

Box teeth, box taste, box eyes, box everything.

If I hear the word "box" again, I think I am going to vomit.

Into The Wild by Jon Krakauer

Two days ago I finally finished reading Jon Krakauer's *Into The Wild.* I read the first five or six chapters at a bookstore in Driggs, Idaho; I finished reading it here at the public library in Dubois, Wyoming. I liked the book a lot. Even though, the death of Chris McCandless was a tragedy, I believe that the two years of his life before his death were redeeming. He experienced more in two years than most people experience in a lifetime. He lived "deliberately" as Henry David Thoreau would have said.

Krakauer writes extensively on his own life and experiences. Krakauer was trying to draw a parallel between his strained relationship with his dad and Chris McCandless' difficult relationship with Walt McCandless. When McCandless found out about his dad's other wife and children, it seemed like he had been living a lie--maybe McCandless felt he was illegitimate: it wounded him deeply. This deep wounding partly drove him into the wild, onto the edge, the fringes of society.

The main reason McCandless hitchhiked, rode freight trains and ended up in the wilderness of Alaska was to prove to himself that he could survive on his own. Krakauer writes of his own mountain climbing experiences; he was young and he wanted to prove to himself that he could climb the mountain and survive some near-death experiences.

At first glance, I thought, how does mountain climbing compare with hitchhiking? Isn't it much more dangerous to climb mountains than to hitchhike? At second glance, people die climbing mountains and people die hitchhiking the highways of the world. Mountain climbers explore and hitchhikers explore: they explore new geographical territory and terrain and they explore their own limits in difficult environments.

McCandless was obviously a very well-read young man. I liked the quotes of various writers at the beginning of each chapter in *Into The Wild.* McCandless left a deep and lasting impression on many people in his travels. Ron Franz, the old guy McCandless met in southern California, was especially touched by his life. I don't see any evidence that McCandless had a relationship with Jesus Christ, but he did believe in God.

When a man of ninety-five dies, people say that he lived a long life and that it was time for him to go. When a young man like McCandless dies at the age of twenty-four, we say it was a tragedy that he died so young. Tragedy is in the eye of the beholder. Yes, I would rather that McCandless had survived his ordeal in the Alaskan wilderness, but he lived more in twenty-four years than some people would live in two hundred years. People have and will learn from McCandless' life and death. It is not how long you live your life, but it is the quality of the life you lived that is important.

People will be reading and writing about McCandless' life for years to come. I saw the film *Into The Wild* for the first time last summer; the cinematography is beautiful--I liked the movie a lot. The hitchhiking scenes in the movie reminded me of my own hitchhiking experiences: the people you meet on the road, sleeping in the desert, the odd jobs you get to make a little money. I may have hitchhiked more miles than McCandless, but he rode more freight trains than I ever will.

I was hitchhiking through Belle Fourche, South Dakota a couple of years ago and this lady picked me up. She told me that she and her boyfriend picked up McCandless while he was hitchhiking through South Dakota back in 1992.

I believe the Lord wanted me to read *Into The Wild* for a reason. There are similarities and differences between my life and McCandless' life. I did a lot of exploratory hitchhiking back in 1986 and 1987, but since 1996, my hitchhiking has been God's will for my life--this is my work: obeying the Lord on the road.

Genesis 47: 9: "And Jacob said unto Pharaoh, The days of my pilgrimage are an hundred and thirty years: few and evil have the days of the years of my life been, and have not attained unto the days of the years of the life of my fathers in the days of their pilgrimage."

Jacob's pilgrimage ended when he was one hundred and forty-seven years old (Genesis 47: 28); Chris McCandless' pilgrimage ended when he was twenty-four; I am still a pilgrim on this earth.

"When the Stranger says: 'What is the meaning of this city?
Do you huddle close together because you love each other?'
What will you answer? 'We all dwell together
To make money from each other'? or 'This is a community'?
And the Stranger will depart and return to the desert.
O my soul, be prepared for the coming of the Stranger,
Be prepared for him who knows how to ask questions."
--T.S. Eliot

Matthew 8: 20: "And Jesus saith unto him, The foxes have holes, and the birds of the air have nests; but the Son of man hath not where to lay his head."

Fixing Fence and the Emigrant Trail

The past couple of days John and I have been fixing some fence and rigging up a water-collecting system for his cow-calf herd. John has some leased ground 35 miles east of Cedarville, California in northwest Nevada.

A year ago I helped John and his wife, Susie, brand over a hundred calves. I have helped work thousands of cattle back in Iowa (I was raised on a cattle farm), but never by roping the calves--we would always run the cattle down a chute. I was glad to have helped John and Susie last year. We should be branding again in a week or so.

This afternoon, after we checked to make sure we had enough water running into two tanks and double-checked the corral fence that we mended, we drove back to the Surprise Valley on the old Emigrant Trail.

I believe the Emigrant Trail was first blazed in the 1840s. This was one of America's first interstate highways. I asked John about the Trail and he said it originally went through Winnemucca and Gerlach, Nevada and then through northwest California and probably into Oregon.

Now that I have ridden down the Emigrant Trail in a rancher's pickup, I can die happy.

*****

[After looking up Emigrant Trail on *Wikipedia*, I believe I traveled on the California Trail (or Applegate Trail) and not the Emigrant Trail.]

The Sunrise This Morning Was Very Beautiful

I am back in Riverton, Wyoming. It took me eight days to hitchhike to Washington, D.C. and nine days to get back. A couple of gas field roughnecks picked me up outside of Rawlins. They were drinking beer and having a good time. They stopped at Sweetwater Station (a bar and grocery store) and had some mixed drinks and played some pool with three other roughnecks. They were getting pretty drunk. The one guy tried to pick a fight with another guy. They were getting fairly loud and obnoxious and finally we headed for Riverton.

I slept on this dirt road just north of Rawlins last night. The air was dry and cool. The sunrise this morning was very beautiful. That is one thing about the desert: spectacular sunrises and sunsets. It is great to be back in dry country: I had some wet clothes in my backpack because I had to walk in the rain when I was going through Iowa. I laid out my wet clothes last night and this morning they were mostly dry. That's the thing about life east of the Missouri River: very humid—hard to stay dry. It is great to be back in Wyoming.

Two Pleasant Surprises: *High Plains Drifter* Revisited

About five days ago I was hitchhiking south of Columbus, Montana and this guy picked me up. He drove me to Absarokee.

I told him that I had been hitchhiking for a number of years. I also said that my book *High Plains Drifter* was published two years ago. He said that a friend of his from Forsyth, Montana saw my book at the public library there and had read it. I never told him that my book was at the library in Forsyth. That was a pleasant surprise.

18 November 2010

This morning I hitchhiked from Victor, Idaho to Wilson, Wyoming. I got dropped off at this gas station in Wilson and I bought a candy bar there. As I walked through Wilson heading towards Jackson, I saw this guy walking to his vehicle that was parked on the shoulder. He looked at me and I looked at him; I thought I recognized him.

He pointed at me and I walked up to him and said something like, "I know you. You're the guy from Scotland." (Actually, I think his dad was from Scotland.)

We shook hands and hugged each other. His name was Ian. Ian had picked me up about a year ago and he took me to his place in Wilson and we had a cup of tea; we had a great talk. I think Ian has picked me up twice coming out of Jackson.

So Ian and I were talking and he said, "I read your book and loved it!"

"What?!" I exclaimed. "Wow, no one has ever said that to me before. Thanks. I hope you got something out of it."

Ian told me that a friend of his went through *Amazon.com* and bought a copy of *High Plains Drifter* for him.

We talked for a while longer and then he said that he and his friend were going up to Teton Pass to go skiing. We shook hands and parted company.

God's timing is always perfect. I could have gotten a ride from Victor to Jackson, but no, the Lord had me dropped off in Wilson instead. It was good to see Ian again. I am guessing I will run into him again in the near future.

Wyoming to Utah

I hitchhiked out of Jackson, Wyoming on the 2nd of October. I got a ride with a college kid named Justus from Daniel to south of Pinedale--near the Jonah Field (gas drilling operations). He took a photo of the two of us standing just off of U.S. 191. Justus was going to school at Montana Tech in Butte where he was studying petroleum engineering.

I got a couple of rides to Rock Springs and walked several miles south of I-80 on U.S. 191 and slept outside on the ground near this pipeline they had just put in--it runs between Vernal, Utah and Rock Springs, Wyoming. I woke up the next morning to the sound of an engine running. I popped my head out of my sleeping bag and this pickup stopped just twenty feet away from me. This guy in a hard hat walked over to me and gave me half a sandwich and a bottle of Vitamin Water. We talked for a short while and then he drove off up the pipeline grade.

I got a ride to Vernal and then I got a couple of rides to Duchesne. The ride to Duchesne was with a guy named Dane. He had picked me up before in Idaho two months ago--he gave me a ride to Bozeman, Montana. We had a great talk.

I got another ride to Price, Utah and there a man and woman, Aaron and Debra, picked me up. They stopped at their place and gave me a two-man tent. I was very grateful. We had a good talk; they were interested in looking at my website. Aaron and Debra drove me several miles south of Price and dropped me off at a rest area.

I walked for a mile or so and got two rides to Moab. I stopped at the hostel in Moab and slept there last night. The next morning I walked south on U.S. 191 for a few miles and got picked up by a lady named Kindy.

Kindy was from Salt Lake City and we had a great conversation about the Gospel. Kindy was a graphic designer, taught a class at the University of Utah and was getting her master's at The Naropa Institute in Boulder, Colorado. She asked me about my life on the road and she asked me to recite my poem "Shiloh." I had a copy in my billfold, so I read it to her as we drove down the road. Kindy really liked "Shiloh;" she said my poem reminded her of Mel Gibson's *The Passion of the Christ*.

Kindy dropped me off in Monticello, Utah where I made a photocopy of my "Shiloh" poem at the library. I soon got a ride from Monticello to Blanding where I went to the library and typed some more stuff on the Internet. From Blanding I hope to make it somewhere in Arizona by sundown. Maybe I can try out my two-man tent tonight.

The Computer, Iowa State University and Jane Smiley

Yesterday I hitchhiked from Jackson to Riverton, Wyoming. I was watching TV at my friend's place yesterday afternoon and I saw an interview on C-SPAN. This editor from *The Washington Post* was talking with Jane Smiley. Dr. Smiley was my professor for three days during the fall semester of 1989 at Iowa State University. Dr. Susan Carlson was pregnant, so several professors from the English Department filled in for her just before and after she gave birth.

Jane Smiley used to teach creative writing at Iowa State from 1981 to 1996. She has had a number of books published. In 1992, she was awarded the Pulitzer Prize in fiction for her novel *A Thousand Acres*. This book was later made into a film; the film was released in 1997.

In the C-SPAN interview, the *Post* editor and Jane Smiley were discussing her most recent book, *THE MAN WHO INVENTED THE COMPUTER: The Biography of John Atanasoff, Digital Pioneer*. Dr. Atanasoff was a physics professor at Iowa State in the 1930s and 40s. He helped invent the Atanasoff-Berry Computer. This was the first automatic electronic digital computer.

If I remember right, Jane Smiley's favorite short story was "The Metamorphosis" by Franz Kafka.

Dubois, Wyoming

I got dropped off here in Dubois, Wyoming earlier this afternoon. I will spend some time here at the library typing some things up and then camp out by the river tonight. God willing, I will head to Riverton tomorrow.

This is a popular saying in Wyoming: "There are two seasons in Wyoming: winter and road construction."

There was some road construction as I walked north out of Jackson this morning. Looks like there will be road construction on U.S. 26 between Moran Junction and Dubois later this month. Without road construction we will be a people no more. Road construction is the basis of a sound economy--transportation of goods and national defense. The Romans built roads all over their empire. At one time, all roads led to Rome. Later, all roads led to London. Now, all roads lead to Dubois, Wyoming.

Road construction is eternal.

*****

Luke 9: 58: "And Jesus said unto him, Foxes have holes, and birds of the air have nests; but the Son of man hath not where to lay his head."

"The Son of man hath not where to lay his head."

Sounds like the Son of man had no certain dwelling place. Reminds me of someone I know very well.

If I have money, I will get a motel room. I made some money working for some friends out west, so a motel room is well within my grasp--till the money runs out.

I have slept in abandoned cars, barns, hay stacks, corn stacks, under bridges, homes under construction, homes under slow deconstruction (abandoned), fields, pastures, city parks and what have you. The Lord helps me find places to sleep: "The earth is the Lord's and the fullness thereof."

I have slept in post offices and in a couple of truck stops. I also have a two-man tent that I use whenever I can. I have stayed in missions and shelters and slept under trees. A hitchhiker has to get his sleep somehow.

I once slept in a pickup near these railroad tracks in a small town in Nebraska. I woke up and walked to U.S. 30 and started thumbing for a ride. A half hour later, some guy walked up to the pickup and drove off with it. I am glad that I didn't sleep in that morning.

I am sure that someday the Lord will let me settle down some place. It doesn't really matter where--I am pretty flexible. It doesn't matter where I lay my head . . . as long as I abide in the powerful Presence of God (Zion).

Zion is my home.

Lance Corporal Chance Phelps, USMC, 1984-2004

Yesterday I visited the grave of Chance Phelps at the cemetery in Dubois, Wyoming. The cemetery is located on a hill overlooking Dubois. This was the information that I got from a gravestone and a grave marker:

Chance Russell Phelps
July 14, 1984-
April 9, 2004
Lance Corporal
U.S. Marine Corps
Bronze Star with Valor
Purple Heart
Operation Iraqi Freedom
KIA, Al Anbar, Iraq

I then walked to the elementary/middle school on the north side of town. There was this little park with a sign that read: "Chance Phelps Community Memorial Park".

In the past year, there was a film about Chance Phelps starring Kevin Bacon: *Taking Chance*. It was a *2009 Sundance Film Festival Award* winner. Kevin Bacon won *The Golden Globe* and *Screen Actor's Guild Award* for his portrayal of Lt. Colonel Michael Strobl. I have yet to see the film; I have heard that it is very good.

I remember well back in 2004 I was hitchhiking in the Dubois-Riverton neighborhood and I got dropped off in some town and I looked at this newspaper. There was a photograph on the front page of the *Casper Star-Tribune*. There was a casket and it had an American flag draped over it. The casket was carrying the body of Chance Phelps and it was being transported to the cemetery with a wagon and a team of horses. You could see the Wind River Mountains in the background.

Freedom is not free. Lance Corporal Phelps sacrificed his life so that others could live free.

Lightning Source UK Ltd.
Milton Keynes UK
UKOW040358210513

210980UK00002B/30/P